Who Decides?

Conflicts of Rights in Health Care

CONTEMPORARY ISSUES IN BIOMEDICINE, ETHICS, AND SOCIETY

WHO DECIDES?
Conflicts of Rights in Health Care

Edited by
Nora K. Bell
University of South Carolina

Humana Press • Clifton, New Jersey

Library of Congress Cataloging in Publication Data

Who decides?

(Contemporary issues in biomedicine, ethics, and society)
Includes bibliographies and index.
1. Medical ethics. 2. Medicine — Decision making. 3. Medical laws and legislation. ✓1. Bell, Nora K.
II. Series.
R724.W54   174'.2   81-83908
ISBN 0-89603-034-2   AACR2

We are grateful to Prentice Hall for permission to reuse some material from Ruth Macklin's books, *Man, Mind, and Morality*: The Ethics of Behavior Control (1982).

© 1982 the HUMANA Press Inc.
Crescent Manor
P.O. Box 2148
Clifton, NJ 07015

Printed in the United States of America.

Contributors

George J. Annas, School of Medicine, Boston University, Boston, Massachusetts

Mila Ann Aroskar, School of Public Health, University of Minnesota, Mayo Memorial Building, Minneapolis, Minnesota

Tom L. Beauchamp, The Kennedy Institute of Ethics, Georgetown University, Washington DC

Nora K. Bell, Department of Philosophy, University of South Carolina, Columbia, South Carolina

James A. Bryan, II, Department of Family Medicine, University of North Carolina, Chapel Hill, North Carolina

Daniel Callahan, The Hastings Center, Institute of Society, Ethics and the Life Sciences, Hastings-on Hudson, New York

H. Tristram Engelhardt, Jr., The Kennedy Institute of Ethics, Georgetown University, Washington, DC

Sally Gadow, Department of Community Health and Family Medicine, College of Medicine, University of Florida, Gainesville, Florida

Samuel Gorovitz, Department of Philosophy, University of Maryland, College Park, Maryland

F. Patrick Hubbard, School of Law, University of South Carolina, Columbia, South Carolina

Kevin Lewis, Department of Religious Studies, University of South Carolina, Columbia, South Carolina

Ruth Macklin, The Hastings Center, Institute of Society, Ethics, and the Life Sciences, Hastings-on Hudson, New York

Guenter B. Risse, Department of the History of Medicine, Center for Health Sciences, University of Wisconsin, Madison, Wisconsin

James L. Stiver, Department of Philosophy, University of South Carolina, Columbia, South Carolina

Thomas S. Szasz, Professor of Psychiatry, SUNY College of Medicine, Syracuse, New York 13210

Jay Wolfson, School of Public Health, University of South Carolina, Columbia, South Carolina

Preface

Many of the demands being voiced for a "humanizing" of health care center on the public's concern that they have some say in determining what happens to the individual in health care institutions.

The essays in this volume address fundamental questions of conflicts of rights and autonomy as they affect four selected, controversial areas in health care ethics: the Limits of Professional Autonomy, Refusing/Withdrawing from Treatment, Electing "Heroic" Measures, and Advancing Reproductive Technology. Each of the topics is addressed in such a way that it includes an examination of the locus of responsibility for ethical decision-making. The topics are not intended to exhaustively review those areas of health care provision where conflicts of rights might be said to be an issue. Rather they constitute an examination of the difficulties so often encountered in these specific contexts that we hope will illuminate similar conflicts in other problem areas by raising the level of the reader's moral awareness.

Many books in bioethics appeal only to a limited audience in spite of the fact that their subject matter is of deep personal concern to everyone. In part, this is true because they are frequently written from the perspective of a single discipline or a single profession. As a result, one is often left with the impression that such a book views the philosophical, historical, and/or theological problems as essentially indifferent to clinical, legal, and/or policy-making problems. Books by philosophers, for example, often engage in seemingly endless theoretical investigation for which clinicians have little or no tolerance, and those written by members of the health professions often make isolated clinical judgments that leave philosophers complaining that important aspects of moral theory have been ignored. It is the special intention of this book, then, to bridge the gulfs separating the various professions by presenting a rich panoply of original essays—drawn and carefully enhanced from lectures presented at a well-received interdisciplinary conference—designed specifically to produce fresh understanding on all sides.

It is evident that there are difficult value decisions to be made in all areas of bioethics, just as it is evident that such value judgments cannot be said to be the decision-making prerogative of any single professional specialty. For these reasons, it is important that nurses and physicians, ethicist and philosophers, public policy makers and social scientists cooperate in working through these issues.

One of the things that distinguishes this volume from others on the subject is that, in spite of their mixed perspectives (the authors include nurses, philosophers, physicians, historians, theologians, lawyers, and public health professionals), the contributors capture this cooperative, interdisciplinary spirit. They write for each other and for a truly mixed audience, including students and concerned laypeople. The essays presuppose only minimal acquaintance with the technical terms of the various disciplines and they reflect the feeling that shared expertise can be of great benefit in examining the competing values confronting members of the health professions.

The present arrangement of the material has been chosen for the following reasons: The importance of any moral debate is perhaps most accurately assessed when given proper historical grounding. Thus, Gunter Risse's essay, "Patients and Their Healers: Historical Studies in Health Care," which explores the changing relationships between doctors, their patients, and the societal setting of their association by focusing on five historical therapeutic relationships, has been selected to open the volume.

The essays dealing with the limits of professional autonomy have been placed next for similar reasons. The largest section in the volume, Part I, addresses issues basic to an understanding of the debate over conflicts of rights in health care. The question—Who Decides?—is not simply a matter of whose choice it is whether or when to use heroic measures, or whether or when to use new reproductive techniques. As the essays here demonstrate, to quote Professor H. T. Engelhardt, "it involves the more fundamental issues of who sets the boundaries for the health professions, of who defines what these professions are, and what the goals are that they should be pursuing." The essays in this section set about to examine the geography of boundaries, "of lines among the professions," to ask whether there is any sense to the way these lines are drawn, to ask for a reappraisal of the goals of medical care, and to ask for the locus of responsibility for human well-being. These essays suggest that we need not feel forced to accept the existing health care framework and its prevailing norms as defining the only conditions under which patients and health professionals can interact.

Having worked through these discussions, the reader can then proceed to Parts II, III, and IV with a more developed sense of what is at stake in specific problem areas. In addition, because some of the problems that are presented in the broader context of Part I appear again in the essays in the remaining sections, later discussions profit from this earlier, preliminary introduction of them.

Each of sections II, III, and IV deals with the question of who decides as it relates to *specific* aspects of the therapeutic encounter. Part I, for example, investigates problems associated with the claim that individuals have the "right" to refuse or withdraw from treatment, specifically, psychiatric treatment. The arguments presented by the authors in this section hinge on examinations of the notions of liberty, autonomy, and competence.

The essays on Part III consider what might be said to count as "heroic," when heroics are relevant, when heroics are morally required, and when one could be said to be morally justified in ceasing heroic forms of treatment. In an attempt to suggest answers to some of these questions, the essays also undertake to examine who, if anyone, can be said to have the right to make such determinations.

The final section deals with a range of ethical problems, those associated with human reproduction, that are said to be "genuinely new and unprecedented." These essays examine the multiplicity of issues involved in the ethics of reproductive biology with a view toward answering questions about the rights of researchers, the rights of prospective parents, the rights of children born as a result of using such techniques, and the role of the law in dealing with such questions.

Obviously, not all aspects of these topics can be addressed within the space limitations of this volume. Each topic by itself has been the subject of endless hours of study and has occupied untold numbers of specialists. Understandably, too, some of the views expressed reflect the biases of the authors' professional specialties. Even so, the volume is intended, and should manage, to provide direction to readers interested in pursuing the problems further. To that end, lists of recommended readings have been provided at the end of the Introduction and at the end of each section to serve as both supplementary and complementary material.

The Introduction is designed to assist the reader in two important ways. First, it seeks to provide a reasonably simple theoretical foundation for an examination of claims about rights, one that avoids technical language and jargon as much as is possible. Second, since no book covers precisely what the reader would like, it attempts to place these issues in a broader context by exploring problems and assumptions common not only to the issues discussed in the essays in this volume but to other, wider social issues as well.

It is a distinct pleasure to thank the contributors for their unhesitating cooperation; Thomas Lanigan of Humana Press for his support, encouragement, and assistance; the South Carolina Committee for the Humanities (an agent of the NEH) for providing partial funding for the conference at which these papers were initially presented; the College of Nursing, the School of Medicine, and the Department of Philosophy at the University of South Carolina for their cooperation in putting on the conference; Stephen Darwall and Ferdinand Schoeman for offering helpful criticisms of an earlier draft of the Introduction; editors of other volumes, among them Tom Regan, Tom Beauchamp, and James Childress, whose own efforts inspired my ideas for the structure and organization of the Introduction; Bonnie Lane and Pat Jones for their assistance with the typing; and my husband, David, and my children, Caroline, Elizabeth, and Thomas, for helping to provide me with the time I needed to see this project to completion.

February, 1982   Nora K. Bell

Contents

Part I: Limits of Professional Autonomy

Part II. Refusing/Withdrawing from Treatment

Part III. Death and Dying: Electing Heroic Measures

Part IV. Advancing Reproductive Technology

Chapter 1

Introduction

Nora K. Bell

The essays in this volume all deal with questions involving conflicts of rights—questions that take on an added importance in health care contexts. A quick scan of media reports and the popular literature soon reveals that health care contexts are fertile ground for generating such conflicts: There is a young boy, "brain dead" and decomposing, whose father (contrary to the advice of all the attending physicians) is attempting to block the removal of the child's mechanical life-support systems. A nurse is suspended from practice, her license revoked for six months, because she gave a patient more information about alternative therapies than the doctor had authorized. A psychiatric patient, "sane" enough to commit herself for treatment and "sane" enough to consent to electroshock therapy, is judged neither sane enough to refuse to be drugged into tranquility nor sane enough to withdraw from the treatment. Institutionalized "mental patients" who can neither read nor write "consent" to sterilization procedures.

How are we to react to such cases? Do physicians have the right to withdraw life-support systems, over a parent's objections, from a boy declared "brain dead"? Does a nurse have the right to ensure that a patient is fully informed when the giving of such information is a flagrant violation of a physician's orders? Or, does a patient's right to full and adequate information include the right to know about therapies not deemed "legitimate" options by the physician? Do we have the right to institutionalize and drug a patient against his or her will because such a patient is diagnosed as being mentally ill? Is someone incompetent to make certain decisions for himself or herself simply in virtue of being a "mental patient"?

Elsewhere, we read of surrogate mothers and surrogate fathers making it possible for infertile couples to reproduce. We hear that geneticists "match up" donated sperm in order to fulfill the desires of prospective parents. We read of adoptive children searching for their biological parents, or of cases where adoptive children have borne genetically defective offspring because they were unaware of their real genetic history. The questions raised by the use of sperm donors and donors of ova, not to mention those raised by the use of surrogate mothers and laboratory conception, seem infinitely more complex. Does a child born of such advanced reproductive techniques have rights that children born in more conventional ways might not have? Do couples who have used donated sperm or ova have the right to pretend full biological parentage? Do donors have the right to remain anonymous if the genetic well-being of future offspring is at stake?

The issues to be faced go beyond those dealing primarily with conflicts between individuals. Also at stake are questions growing out of conflicts between society's interests and the individual's interests. For example, what right does one have to prolonged use of exotic life-saving therapies when a larger number of individuals might be successfully treated with the same expenditure of resources? Or, what of the patient in psychotherapy who discloses that he would like to, or intends to, kill several people? Or, what right does one have to avail herself of new reproductive therapies when the threat of overpopulation suggests a need to limit reproduction? And so on.

At yet another level we are faced with interprofessional conflicts. If the various health professions are viewed as being autonomous, then what are the limits of the authority of each? Who establishes these boundaries? Does a nurse have a right to make treatment decisions for a patient? Does a nurse have a right to refuse to follow a doctor's orders if the order is judged to be contrary to the patient's well-being? Does a pharmacist have the right to dispense the less costly generic equivalent of a drug without the express consent of the prescribing physician? Does a physician have a *right to a patient* that makes the physician the primary or dominant member of the health care team?

The questions come all too easily, but the disposition of such cases, the answers to such questions are much more difficult—in part, because they threaten ultimately to have an effect on the well-being of each of us.

The authors of the essays in this volume have carefully examined some of these very same questions. Each essay is an attempt at understanding questions like those posed above as well as an attempt at offering a reasoned solution to some of the problems. However, because the authors represent diverse disciplines and widely divergent viewpoints, their approaches to the problems are quite mixed. Some are more medically oriented than others, some more legal and/or public policy oriented. Some are more philosophical in their approach, some more theological. Therefore, in order to assist the reader in acquiring a basic theoretical foundation useful for an examination of claims about rights, this Introduction will seek to explain general concepts and ideas that are essential to an understanding of the problems no matter what the author's orientation. In that way the essays can be viewed in a broader perspective and the reader will be able to identify what assumptions and biases they have in common.

In discussions of health care and the ethics of health care, few words are used as loosely as is "right". We hear assertions about the right to life, the right to die with dignity, the right to determine what is to be done to one's own body, patients' rights, the right to health care, and so on. For that reason, it is important to get some clear sense of what is being claimed with each of these "rights".

In particular, before we can understand what is at stake in the essays in this volume we need an explanation of what is meant by the word "right", what sorts of beings possess rights, what relationship (if any) exists between rights and responsibilities, what happens when rights conflict, whether rights can be forfeited or overridden, and how talk about rights can settle the question of what ought to be done in particular cases.

What Are Rights?*

Rights language functions in such a way as to suggest that a right is something that is both desirable to have and beneficial to its pos-

*My treatment of rights and the relationship between rights and responsibilities owes a great deal to (1) H. L. A. Hart, "Are There Any Natural Rights?", (2) Joel Feinberg, *Social Philosophy*, (3) W. Hohfeld, *Fundamental Legal Conceptions*, and (4) L. C. Becker, "Three Types of Rights." (See bibliographical data below.)

sessor. But, if one claims to have a right, he or she does not view its possession as something for which the proper response is gratitude or as something that depends for its existence on another's beneficence or good will. Rather, rights language suggests that a right is something to be demanded either as one's own or as one's due. But how is one to understand what a right is?

Various analyses of the concept of a right have suggested answers that generally alternate between asserting that rights are an individual's *entitlements* to be treated in certain ways and asserting that rights are special *claims* individuals can make, or have made on their behalf, either to certain things or to the performance of certain actions. In other words, when we have rights, some say, we are *entitled* to expect others to act accordingly. Or, as others say, we make a demand or *claim* that others are obligated to honor or to which they are obligated to accede.

But, as Hohfeld and others following him have suggested, to characterize rights as either claims or entitlements is too simplistic an analysis. True, rights exist in certain sorts of "rights-relationships," and, true, one may have rights that in the strict sense are "*claims* against someone for an act or a forbearance," but not only are some rights thought to be claims that are indistinguishable from entitlements (some argue that children's rights are such), rights may be further characterized as *privileges* (liberty rights, for example) that are neither claims nor entitlements and that correlate simply with the absence of a claim (a "no-right") in others. Or, as Hohfeld continues, some rights are appropriately characterized as *powers* one has (such as the right to sell his house) that correlate with liabilities in others; and still other rights are called *immunities* (such as the right to remain silent) that correlate with disabilities in others. Interestingly, what all of these analyses share is the notion that having a right involves having a "right-correlative" (to use another Hohfeldian term)—a moral justification for constraining, restricting, interfering with, or determining the actions of another. Of course, in the case of liberty rights, the right-correlative is the *absence* of such a justification.

So, for example, put one way: if Mary has a right to X, then, in pursuit of X, Mary has a justification for interfering with another's freedom that other individuals may *not* have. Mary may be said to have a right, for example, to collect what John promised to pay her. Mary's right to be paid what John promised makes it morally ligitimate for her to determine how John shall act *in this respect*. That is, her right constitutes a morally justified limitation on John's freedom to do with his money as he chooses. But some other individual, say Sally, cannot with any justification make that demand of John if John made no such promise to Sally, unless, of course, Sally is making the demand on Mary's behalf.

Viewed another way, if Mary has a right to X, then others have no justification for interfering with Mary's pursuit or possession of X, so long as in the exercise of her right Mary does not infringe rights others may have. If, for example, Mary has the right to worship as she pleases, then others are constrained from interfering with the practice of her religion so long as Mary's manner of worship does not conflict with others' rights or so long as Mary's manner of worship correlates with a "no-right" in others. In other words, Mary would have grounds for objecting to another's interference as *having no justification*.

This last consideration is of some importance to Bryan's and Beauchamp's discussions, especially concerning cases where a dying patient's desire to avoid the utilization of "heroic" measures conflicts with the desires of family members or physician to implement such measures (or vice versa). One's death might be said to be uniquely one's own. When one is dying, the right to determine the manner in which he or she dies may preclude interference by family or physician. It may be, given the above analysis, that there is no moral justification for interfering.

Negative and Positive Rights

Historically, the conceptual treatment of rights has emphasized a distinction between rights that serve to prohibit others from interfering in our conduct of our lives (sometimes called "negative rights") and those that seem to require others to do or provide certain things for us (sometimes called "positive rights"). The rights of noninterference include such things as the right not to be denied employment on the basis of sex, race, or religion, the right not to be enslaved, and the rights of life and liberty. The positive rights, sometimes also called rights of opportunity, include such things as the right to education, the right to food and shelter, and the right to high quality health care.

With the so-called positive rights, however, not only is it often quite difficult to determine against whom one has such a right, but the obligations said to be generated by such rights are far from clear cut. If we were to grant that there is a right to high quality medical care, for example, can the US government in a time of scarcity be said to have violated that right by failing to ensure that the poor get medical treatment equivalent to that of the wealthier members of our society? Or do physicians violate that right when they refuse to take new patients because their practices are too full? Have hospitals violated that right in not alleviating the shortages of nursing personnel? These positive rights are so open-ended that it is not only difficult to determine whether and when one has been violated, but, more importantly, how far they require individuals and institutions to go.

The rights of noninterference, on the other hand, are slightly less difficult to deal with. For the most part, rights of noninterference appear to be derivative from what most theorists call "the equal right of all persons to be free," that is, rights derived from the notion that one be permitted the liberty to do anything that does not harm, coerce, or restrain others, or in another formulation, rights that correlate with the absence of a (claim) right in others. So, for example, the right to freely exercise one's religious beliefs establishes an obligation of noninterference upon others, provided that the practice of one's religion neither harms nor coerces (nor conflicts with the rights of) another. It is for these sorts of reasons that the courts have been disposed to allow Jehovah's Witnesses to refuse blood transfusions *for themselves* in life and death situations, but not for their minor children. Interference is deemed justifiable and not a violation of one's right when the exercise of that right threatens another.

This distinction poses some difficulties for us though. First, some rights are not so readily categorized as one or the other. The categories are not mutually exclusive. For example, what does one call the right to die with dignity? A right of noninterference, a negative right? Or is it a positive right requiring someone to take positive steps to end the inhumanity of a torturous death? Depending upon how this question is answered, one may find oneself committed to a particular view on the moral acceptability of euthanasia.

The distinction poses further difficulties because there are occassions when the rights of noninterference are said to be meaningless without some complementary positive rights. Consider the Supreme Court's decision in *Roe vs Wade* that a woman has a right to an abortion. The Court's decision affirmed one's right to privacy (a right derivative from something like the equal right of all persons to be free), that is, one's right to determine what shall be done with (or to) her own body. Until the final trimester, a pregnant woman is guaranteed absence of interference should she choose to have an abortion.

The furor surrounding the Court's decision that the government is nevertheless *not responsible* for funding abortions for the poor was generated precisely because some feel that there is a fundamental connection between the having of such a right and an obligation on the part of some individual or institution to provide the framework within which, or the means by which, one can exercise that right. They want to know what it could possibly mean to say that a poor woman has a right to an abortion if she clearly cannot afford it, if for financial reasons she is denied access to skilled medical personnel and adequate health care facilities.

In wanting to argue that poverty *ought not* to limit a woman's right to abortion, in feeling the pressure to admit that one *ought* to be afforded the avenues by which to exercise such a right, in short, in asking a question like the one above, one is making an appeal to *considerations of justice.*

In other words, it may be true that the right not to be denied an abortion is recognized if the procedure is legalized. One is then free to exercise that right if she is a female pregnant with an unwanted fetus. But the *worth of her freedom*, the value of that right (to borrow a Rawlsian notion) then becomes a matter of justice. The complaint that poor women (unable to afford an abortion) want to lodge may be something like this: "I could never acquire the education necessary for developing my capacities to a level that would make me competitive on the job market because good public schooling was geared to the already advantaged in our society." Or, "Society's economic barriers were such that I could never acquire the wherewithal to afford *any* surgical procedure." Such complaints hinge on the social and economic aspects of a system that determine whether or not individuals *can* exercise their rights.

Accordingly, it may be the responsibility of a just society to ensure the value of one's rights by guaranteeing something like a minimum level of income and education. So, in other words, if one has a right to an abortion, then the worth of that right is guaranteed by a just structuring of the social and economic institutions of society (whatever one takes justice to require).

On this analysis, "rights of opportunity," at least insofar as they are thought to complement rights of noninterference, are no more than responsibilities required by justice in assuring the worth of one's liberty. At the same time, however, it must also be acknowledged that there is a limit to the responsibility to provide assurances for the implementation of such rights. It may *not* be the business of justice, for example, to fund an abortion for a woman if she refuses birth control assistance from her department of public health, or if she refuses to use birth control measures yet remains sexually active (necessitating repeated abortions). The difficult question, of course, is that of determining just what the limit is.

Moral and Legal Rights

Rights that are asserted to exist as a result of governmental guarantees, clearly public laws, and/or intitutional regulations are called *legal rights*. Those that are said to exist prior to and independent of such intitutional guarantees and prescriptions are called *moral rights*. That people are "endowed by their Creator with certain unalienable rights; that among these are life, liberty, and the pursuit of happiness" is an expression of what many take to be among our basic moral rights. Still others include as additional moral rights such things as the right to die with dignity, the right to bear children, the right to health care, the right to control what happens to one's body, and so on. On the other hand, our legal rights are understood to include such things as our Constitutional guarantees: the right to a jury trial, the right to exercise religious freedom, the right to assemble peaceably, the right to vote, and so on.

Most rights theorists want to maintain that moral and legal rights differ in important respects: Moral rights are thought to be universal, at least in the sense that at some level of analysis they are thought to accord with a universal principle. That is, if something is a moral right, say, the right to die with dignity, then there is a sense in which we want to say that all people no matter where they live have this same moral right whether or not it is formally recognized as a legal right in that particular location. Legal rights, on the other hand, may (and do) vary both from time to time and from location to location. One's right to purchase and consume alcoholic beverages, one's right to do business on Sunday, one's right to drive a car all change, for example, from region to region within the United States and, of course, elsewhere as well. In addition, there are legal rights possessed by the citizens of some countries, such as the right to a trial by jury and the right to vote, that are not rights of citizens of other countries.

Moral rights are further distinguished from legal rights by being regarded as inalienable and equally held. So, whereas one may have certain legal rights taken away, such as when a convicted felon loses the right to vote, whatever moral rights one is said to possess can neither be taken away by nor given over to another. Furthermore, whereas legal rights are held by some and not others, e.g., only those eighteen years of age and older have the right to vote, moral rights cannot be held to a greater extent by some (say, males) than others (say, females).

Generally speaking, many people view moral rights as the basis for the establishment of our legal rights, and if they feel that they have this or that moral right they may then look to the government or to institutions to give it a formalized legal status. So, for example, some women (and men) have pushed for ratification of the Equal Rights Amendment as a means of accomplishing a formal recognition of what they view as the moral right of equality. Similarly, many of the so-called "rights of children" or "rights of the unborn" against practices associated with artificial insemination and artificial inovulation are thought to be moral rights that *ought* to be formalized, *ought* to be made the law. Annas' argument that the procedures for the selection of sperm donors and AID record-keeping practices should be set up so as to maximize the best interests (or rights) of the child is an appeal to just such a notion. The theme emerges again in Callahan's essay when he discusses the view that reproductive techniques such as *in vitro* fertilization might present a hazard to the fetus —that the possibility of its bringing harm to the unborn warrants formal intervention to bring about a moratorium on the use of such techniques. Again, the feeling is that there are moral rights at stake here that *ought* to be recognized by the law.

But of course the connection is not that easy to make. Clearly, many things that are (or might be) legal are not necessarily moral, and certainly many things that are moral rights need not and ought not be made legal. If I promise to look after your ailing dog while you are on vacation, then surely you have a right to expect that I will, just as I have an obligation to live up to my word. If you return to find that your dog has died of my neglect, then I have violated your right and failed in my obligation.

The distinction made above suggests that your right was a moral right, just as my obligation was a moral one. But to ask that laws be passed to govern such a promise-keeping situation is going beyond what we want the law to do. However, it is important to see that it is unlikely that we would say that the force of the obligation is any less because it is not "the law." It is still binding, though *morally* binding. Conversely, we would not be likely to fulfill a legal obligation (if we consider it to be immoral) simply because it is "the law."

Rights and Responsibilities

The notion of having a right carries with it the notion of responsilibity or obligation (a "right-correlative"), but the nature of the relationship between rights and responsibility warrants closer examination.

If, as was discussed earlier, Mary has a right to X, then the rest of us have an obligation not to interfere with Mary's pursuit or possession of X, provided we are not thereby harmed. In other words, an individual's possession of a right makes others responsible for acknowledging it and permitting its exercise.

But does Mary's right to X place Mary under any special obligations? Is Mary allowed to exercise that right unconditionally? If Mary has the right to life, for example, does that mean that she can take my medicine if that medicine is all the hope either of us has for recovery? If John has told his psychiatrist that he intends to use violence against several different people and if John has the right to refuse or withdraw from treatment, does that mean that John should be permitted his freedom? What some want to say of such cases is that Mary's and John's rights impose certain responsibilities on them as well, at least *one* of which is to avoid unjustly harming others in the exercise of those rights. It might be more accurate to say that these reponsibilities do not arise because of the rights Mary and John have, rather the responsibility to avoid unjust harm to others is generated by *others'* rights to certain things. In any case, rights must certainly be limited in this sense.

Of course, appeal to the "harm" criterion does not dispense with all the difficulties here. It has been a matter of considerable debate whether the exercise of certain rights has brought unjust harm to others. For example, it has been argued that a woman's right to the control of her own body, her freedom to choose elective abortion, unjustly harms the fetus. Advanced reproductive techniques, though affirming one's rights to procreate, might arguably be said to bring harm to the children who must bear the stigma of having been born by such "unnatural" means. Whether the prohibition against harm constitutes adequate justification for the violation of one's rights is discussed in a number of the essays, those by Callahan and Annas in particular, as well as those by Szasz, Macklin, and Stiver.

What is interesting is that appeal to the "harm" criterion is sometimes expanded to include cases where, as Szasz points out, one might be about to harm oneself. That is, one is held reponsible for exercising the right in such a way that one neither harms anyone else *nor harms oneself*. But there is a paradox involved in this: If a person *voluntarily* chooses to perform an action that is harmful to oneself, this is taken to be a sign of an inability to act responsibly; the act is regarded as *not voluntary*—a sign that one is not "competent" to exercise autonomous choice. In such cases, some argue, it is justifiable to violate that individual's rights. And it is in precisely such cases, that the question, "Who decides?", deserves special attention.

Although it may be accurate to say that rights against unjustified interference are thought of as rights imposing obligations on everyone, there are many instances where we speak of rights and their corresponding obligations in a much more limited sense. One of the more fundamental relationships between rights and responsibilities becomes evident when one considers the nature of various transactions that occur between individuals. A patient, for example, who is offered a much sought-after berth in a weight control clinic on the condition that the patient walk fifteen miles a day is party to a double transaction. Once the patient voluntarily chooses to become a participant in the program, that patient can be said to have a right to expect those services set out as a part of the program. Similarly, the clinic itself, in accepting the patient under those conditions, has assumed obligations for that person's treatment in the ways the program format prescribes. In addition, however, the clinic has a right to expect the patient to live up to the terms of that patient's acceptance into the program, just as the patient has the obligation to do what is required of a participant. In other words, rights and obligations may be created by mutual submission to certain conditions or restrictions.

On such an analysis, rights are viewed as generated by "rules" of various kinds—rules of custom, formal legal regulations, covenants, mutual practices, even rules of games. What is important about such rights is that they grow out of special *voluntary* transactions between individuals as well as out of special relationships in which individuals stand to one another. In other words, rights are viewed as growing out of a formal or informal systematizing of various practices whereby individuals voluntarily incur obligations and generate rights, and the "rules" define rights and obligations bilaterally.

Thus, if a patient freely chooses to submit to a physician's care in matters concerning health, then perhaps, as Gorovitz suggests, the patient's health is no longer solely the patient's business. The patient has accorded the physician certain rights by authorizing that physician to interfere with the patient's health in certain ways, just as, in accepting the authorization, the physician has assumed certain obligations. It may be the nature of the doctor/patient relationship, the nature of "doctoring," that doctors cannot mind their own business. By the same token, if a patient who is fully informed of the risks or discomfort associated with a particular procedure freely chooses to go ahead with it, then perhaps it is appropriate to say that he or she has no right to seek redress if the procedure is not pleasant or successful.

Another implication of claiming this bilateral relationship between rights and responsibilities is that, because they are mutually restrictive, it seems perfectly permissable for an individual's right to be denied, especially in conditions of scarcity, if that individual refuses to fulfill an obligation assumed under the rules. Thus, if the weight control patient mentioned above were to refuse to walk the required fifteen miles a day, it would hardly seem improper to deny that person continued treatment in the program. *Both* the patient and the clinic were parties to the transaction, but, inasmuch as the patient broke the context in which the clinic's duty was binding, perhaps the clinic's obligation to provide services ceases.

There are two points to be made here: one is that such rights can be limited by conditions external to the transaction, e.g., scarcity, such that agreeing to provide certain services cannot obligate one unconditionally. The other is that responsibility for the maintenance of such a right is not one-sided. An individual's claim to a right in such cases holds only so long as that person does what is called for by the rules. Wolfson argues, for example, that one's right to health care may be limited by the requirement that one exercise responsibility for one's own health in specified ways. If one smokes, for example, then in conditions of scarcity, one may be limited in the kinds of health services for which one is eligible; i.e., that person may be denied treatment for emphysema or chronic bronchitis.

The paradigm for such an analysis is that of entering into a contract, where the parties to the contract are considered to be autonomous moral agents. Such a view of the relationship between rights and responsibilities underlies the stance Gadow adopts in her essay. She develops a self-care philosophy in which health professionals and physicians are viewed as offering positive assistance to individuals interested in developing and exercising autonomy in health matters. The key to such a self-care philosophy is the emphasis placed on autonomy and mutual cooperation. The individual freely chooses certain therapies and actions in the interest of promoting his or her own health—one's rights and responsibilities with respect to his or her own health are derivative from one's autonomy.

Who Has These Rights?

If the notion of a moral right is clear, if the relationship between rights and responsibility is clear, if it is clear that responsibility involves the notion of conscious, voluntary choice, it then becomes important to discern *which* beings have these rights, who are responsible beings. Do comatose individuals have such rights? Do psychiatric patients? Do fetuses? Do children as yet unconceived? Do infants?

Consider some of the criteria frequently offered for determining who or what possesses such rights:

1. That one be capable of autonomous action, that is, of free, rational self-direction.
2. That one have an enduring concept of self.
3. That one be a language-user.
4. That one be conscious.
5. That one be a person.

It is undoubtedly immediately apparent that each of these criteria, considered singly and jointly, rules out beings that we would want to say *do* possess rights. If, for example, having an enduring concept of self is a necessary condition for possessing rights, then infants, individuals who are comatose, and fetuses (to name a few) would have none of these rights. Acceptance of such a criterion would rule out any discussion of the rights of the unborn or rights of the unconceived. Insofar as arguments like those offered by George Annas for regulations governing artificial reproductive techniques so as to protect "the best interests of the (yet unborn) child" hinge on there being something like rights of the unborn, this criterion would be unacceptable (as would several of the others).

Or if we want to say that the mentally ill, the severely retarded, the senile or those of "diminished autonomy" have rights, then being capable of free and rational self-direction cannot be a necessary condition for possessing rights, nor can being a language user. When Macklin argues, for example, that paternalism may be justified in cases when the individual is temporarily incapable of "self-legislation", she does not argue that such individuals no longer possess certain rights. Rather, she argues that to allow them to *exercise* certain of their rights, namely, the right to refuse treatment, may be to harm them unduly by preventing their return to more complete autonomy. In other words, they still possess the right(s), but are incapable at the moment of exercising the right(s) properly.

That one be conscious and that one be a person are criteria no less difficult to deal with. The criterion of being conscious, for example, would allow us to include perhaps more beings than some want included—namely, animals that are not human—and if such nonhuman animals have rights then not only does "health care" take on new meaning, but so does "the ethics of health care." Or it may exclude beings that some want included—namely, human fetuses. The criterion of being a person is equally worrisome, for it introduces all the philosophical difficulties inherent in attempts at setting forth criteria for personhood, some of which are the very same criteria offered for determining possession of rights.

There is the further difficulty that the problem of setting out those conditions that beings must satisfy in order to have rights seems intimately connected with problems of determining competence and noncompetence. That is, many of the same criteria are suggested as bases for determining patient competence and some argue that when a patient fails to measure up to standards of competence, the patient also fails to qualify as having certain rights. Others argue, however, that failing to measure up to standards of competence does not do away with a person's rights, but it does justify interfering with that individual's liberty. So, for example, one's threatening to do grave harm to oneself is judged an irrational act, a sign of diminished autonomy, and hence, ample justification for interfering with the right to control what happens to that person's body.

In spite of the above difficulties, the debate surrounding how to determine the criteria essential for possessing a right is not a frivolous one. If comatose individuals *do* have certain rights, then what justification can be offered for ceasing "heroic" efforts? Does anyone else have the right to decide that heroics are no longer appropriate in treating such an individual? As the reader works through these essays it will become apparent that each of the authors has had to come to grips with the problem of deciding what sorts of beings have rights. Just as Annas would have to believe that the "potential for becoming a person" accords rights, so Szasz and Macklin would seem committed to the view that the "potential for free and rational self-directedness" qualifies one for possession of rights. But deciding which beings have rights is only part of the problem, it is also important to examine, the bases for rights claims, that is, the grounds offered for justifying the claim that someone or something has a right.

Some want to argue, for example, that there are rights that do *not* depend (in principle) upon the possessor's having conscious interests, making claims upon others, or making agreements with others. Such rights depend only upon whether there is a duty toward the putative right-holder. These rights are called *derivative* rights, i.e., rights derived from the logically prior existence of a duty. Derivative rights, inasmuch as they are generated by the existence of duties in others, may be held by anyone or anything to whom duties are owed. So, at least in principle, such rights may be held by trees, rocks, nonhuman animals, human fetuses, plants, the permanently comatose, etc. But if one were to grant the above analysis of derivative rights, it *does not* follow that inanimate objects, nonhuman animals, and other forms of nonhuman life *do have rights*, for whether or not there are in fact *duties* to these sorts of beings is not settled by such an analysis.

A final comment about possessing rights seems in order: it will be clear from reading Aroskar's and Gorovitz's essays that rights are sometimes claimed by individuals in virtue of certain roles they have assumed. For example, one who becomes a nurse has certain rights (and responsibilities) *as a nurse* that anyone else becoming a nurse also has. Similarly, one who becomes a physician has certain rights and responsibilities *as a physician* that others (nonphysicians) do not have. In other words, the possession of rights is not limited to beings; other entities, i.e., corporations, professions, and so on, also have rights.

When Rights Conflict

Merely establishing that there are instances when individuals can properly claim to have rights and that those rights impose correlative obligations on others is not enough. One wants to know the scope of such rights. It is not difficult to think of cases where it would be appropiate to say, "That person may have the right to do X, but ought not to do it." Or, "She is obligated to do X, but she ought not to do it." For example, a physician who has a form signed by the patient giving consent to a radical mastectomy, even though the patient explicitly refused to undergo anything other than removal of a lump, has the right to perform the mastectomy though he ought not to do it. Or consider the soldier who has taken an oath of obedience to his superiors. A superior officer orders the soldier to shoot all infants under two years of age. The soldier is obligated to obey the officer, but ought not to do it.

Similarly, we can think of cases where we might be temped to say that an individual "forfeits" right(s) by certain kinds of actions. (Some might say that the weight control patient who refuses to walk the allotted fifteen miles "forfeits" the right to treatment.) It is important, in other words, to know whether limitations are applicable and whether other moral obligations in certain circumstances might sometimes override a right.

Clearly, when we assert that all persons have the "inalienable" rights of life, liberty, and the pursuit of happiness we intend that it be understood that such rights are limited in scope. To call moral rights universal and inalienable is not to say that they are absolute and unlimited. In other words, the right to liberty is not an absolute right; it does not entitle us to infringe another's right to liberty in pursuit of our own interests. It does not entitle us to murder or steal or plunder at will. As discussed above, it must yield to the limitation of avoiding unjust coercion or harm to others.

For a right to be absolute, it must in all circumstances and for all instances be undeniable. For any moral right to be absolute in that sense would mean that in cases of conflict other rights would have to give way to this one. Similarly, these other rights, if they must be limited or must give way, cannot themselves be absolute. Clearly, one's right to liberty is not absolute in this sense, for it *does* conflict with other rights, namely, another's right not to be coerced or unjustly harmed, and in this respect the right to liberty is said to be limited. The same is true of the many "human rights" delineated in the Universal Declaration of Human Rights passed by the United Nations General Assembly in 1948. The right to health care, for example, cannot be an absolute right, for if it were it would mean that no cost could be spared in providing health care to all in need of it; it would mean that our rights to education, food, shelter, clothing, etc. would have to be compromised in cases of conflict. These human rights, too, must be limited—in many cases by the requirements for the satisfaction of other of our rights.

Equally clearly, I think, the special rights and obligations discussed above must also be understood as limited in scope. I may promise a friend to do something on an appointed day at an appointed hour, thereby incurring the obligation to act as I have promised. However, when the day and hour arrive, suppose that I am called to attend to my child, taken ill quite suddenly at school. In seeing to my child, I break my promise to my friend.

Likewise, one who takes the Hippocratic Oath vows not to give any woman "an abortive remedy." Yet, suppose that the patient's life is in jeopardy unless the developing embryo is aborted. In acting to save the patient's life, that is, in performing this abortion, the physician breaks a solemn vow.

But an obligation generated by a voluntary transaction such as promising or the taking of an oath cannot be viewed as binding *at all costs*. However, the force of the obligation generated by my promise is no less, given my child's illness, nor would we say that my child's illness revokes or nullifies my friend's right to expect the promised action. By the same token, the obligation generated by the physician's oath is no less, given the threat to a patient's life. Such moral intuition is evidenced, I think, by the need one feels for explaining the circumstances surrounding the breaking of the promise or the oath. Generally, in such cases one pleads an *overriding moral obligation*, that is, one that justifiably limits a conflicting obligation.

If, as just suggested, the notion of overriding moral obligations does not nullify or revoke rights individuals may have, but merely restricts them to the extent warranted by the obligation, then one could argue (much to the dismay of followers of Szasz) that institutionalizing paranoid schizophrenics, for example, does not deny their right to liberty (only to a certain measure of liberty) so much as it asserts an overriding obligation to protect the average citizen. Or one could argue, and I think Bryan and Beauchamp would agree, that foregoing heroics in cases involving dying patients does not deny one's right to life, rather it affirms an overriding moral obligation to allow one to die. Similarly, continuing the use of advanced reproductive techniques does not deny the rights of the unborn as much as it affirms an overriding right of the already born to bear children.

One can also put this in terms of the violate/infringe distinction. That is, insofar as one can plead an overriding obligation, the interference with another's right is justified and merely *infringes* that right. An unjustified interference with another's right, however, is said to *violate* that right.

In other words, just as the existence of special rights and obligations generated by oaths and promises does not entail that there are no cases where breaking one's promise or oath will be morally justified, just as mutual submission to social, political, legal, and/or institutional restrictions does not entail that there will be no cases in which ignoring the restrictions is morally justified, so it is with the rights and obligations that concern us in health care contexts. There will be cases when such rights may be infringed. There are disputes and conflicts and there is often the need to determine overriding moral obligation.

Consider, for example, the disputes that arise involving children of members of the Jehovah's Witnesses sect. For religious reasons, even though it may mean the death of the child, Jehovah's Witnesses parents refuse to permit physicians to administer blood transfusions. The parents, of course, argue that giving the transfusions would violate not only their right to religious freedom, but also rights they have as parents. When physicians ask the courts for permission to give transfusions to such a child, they claim that the obligation to save the child's life overrides any obligation they may have to respect the religious freedom or the familial rights of the parents.

Or consider cases of individuals who voluntarily choose to do things injurious to their health. Some want to argue that the obligation to use our medical resources wisely overrides the right to health care of those who refuse to exercise some degree of responsibility with respect of their own health. In other words, not only do rights conflict with other rights, rights also conflict with considerations of the general welfare.

The real difficulty in all such cases comes in attempting to determine what has sufficient moral weight to count as an overriding moral obligation. Do we want to argue, following Ronald Dworkin, that rights "trump" all other kinds of considerations?

Standards of judgment vary and certainly have not eliminated the controversy surrounding such decisions. Such standards (rightly or wrongly) include the following:

Importance of Autonomy

There is something very important about one's having a feeling of autonomy, or self-determination, of control over one's destiny. In the discussions in the essays by Szasz, Macklin, and Stiver, for example, the dispute is not over whether a respect for autonomy is morally important. Rather, since autonomy is viewed as conceptually linked to the notion of liberty, the disagreement concerns whether other considerations, such as a diagnosis of "mental illness," can be allowed to override one's right to autonomy.

In a very real sense, the questions about autonomy lie at the heart of the question, "Who decides." If one is competent to decide certain things for oneself, then certain models of professional/patient relationships appear to be unsatisfactory: among them, as Szasz and Stiver would agree, "paternalistic" or "priestly" models of interacting. In other words, no matter how much a health professional feels obligated to do what is "in the patient's best interests," there may be an overriding obligation to acknowledge the patients' right to choose for themselves. As Gadow puts it, a health professional is obligated to acknowledge what is "freely chosen by an individual in the interest of promoting his or her health *as he or she contrues it*" (my emphasis).

On the other hand, if one is not competent to make his or her own decisions, it is important to the ethics of patient care to examine who should make the decisions in these cases.

Although the question of whether one is or is not competent to exercise autonomy is crucial to many aspects of health care ethics, and although this is a question addressed both directly and indirectly in many of the essays in this volume, being recognized as autonomous is not always considered a necessary condition for possessing certain rights, nor does judging one to be autonomous always swing the moral weight in favor of that individual. In other words, establishing a patient's competence or autonomy (or lack thereof) does not resolve all the disputes. The more central question remains that of how to resolve *conflicts of rights*.

For example, as autonomous moral agents prospective parents may have the right to avail themselves of the technique of *in vitro* fertilization. And a Physician, equally possessed of the right to autonomous action, may have the right to offer the procedure to this particular couple. But do these rights conflict with the right that the children born of such a process have to a certain quality of life? That is, do the unborn have the right not to be born into certain kinds of environments, say a family with more children than it can comfortably feed or a society where many starve because of overpopulation, or a culture that looks upon children produced by means of such techniques as freaks? Can the obligation to respect the autonomy of the parents and the physician override such rights? These and similar questions have captured the attention of the authors in Part IV of the volume.

Or suppose that a dying patient who is judged to be competent requests a nurse or physician to administer a lethal dose of pain medication in order to end the patient's suffering. Does the obligation to respect autonomy override the health professional's right to refuse active assistance in the termination of life? In short, the patient may well have the "right to die" as he or she chooses, but there seems nothing morally amiss if we deny that such a right obligates anyone else to help implement it.

The essays in the first section of the volume, in particular that by Aroskar, emphasize yet another aspect of the importance of autonomy that makes conflict decisions even more difficult. Not only is the concept of autonomy important to the individual, but the autonomy of the various health professions is frequently at stake in health care decisions. In their roles as health professionals, nurses, physicians, and public health professionals all have a primary obligation to the patient. If each profession functions autonomously vis-a-vis the patient, and the patient functions autonomously as well, does this not ensure conflicts?

Suppose, for example, that a physician orders pain medication for a dying patient to be given at four-hour intervals. Suppose further that the patient requests additional medication and the nurse sees that the patient needs remedicating at two-hour intervals, yet the physician refuses to alter the original order. Do nurses, as professionals, have the autonomy to countermand the physician's orders? Do they have the professional autonomy to make treatment decisions for patients?

The notion of informed consent is viewed as derived from the concept of autonomy, that is, one can exercise control of one's health or with respect to health) only is one is fully and adequately informed. All health professionals are committed to the concept of informed consent. But suppose a physician fails to inform a patient either of risks inherent in a particular procedure or of less risky alternatives. Does a nurse have the professional autonomy to provide the information without first consulting the physician? Does a physician's autonomy in making treatment decisions override the patient's right to be adequately informed? Does a nurse's autonomy override her obligation to cooperate with other members of a health care team?

As both Gadow and Aroskar emphasize, the notion of professional autonomy does not entail that the professions be at loggerheads. The exercise of one's autonomy does not entail the violation of another's autonomy. Both authors see the health professional's role as one of cooperating in assisting the individual to acquire and exercise more autonomy with respect to matters of health. In other words, they view the obligation to respect the autonomy of the various professions as giving way to an overriding obligation to provide for the patient's welfare, to provide for conditions that will enhance the *patient's* autonomy.

Best Interests of the Patient

Sometimes, even though a patient is judged competent, the obligation to respect autonomy is overridden by a judgment that a patient's choice is not in his or her best interests, and hence, not "truly" autonomous.

For example, laetrile, though touted by some as an effective anti-cancer drug, is deemed worthless by the vast majority of practicing physicians and medical researchers. Some patients, knowledgeable of the controversy over laetrile, yet unwilling to undergo more conventional cancer therapies, assert that their right to control their own destinies entails that they have the right to choose their own form of treatment. Hence, not only do they have a right to use laetrile, they have a right to expect that it be made available to them. However, the medical profession has argued that it has an obligation to provide only good quality medical care to patients and that such an obligation includes rejecting unproven and questionable therapies as not in patients' best interests. In the case of laetrile they are arguing not only that laetrile should not be made available, but that doctors should not prescribe it even if patients want it. In other words, when a treatment or technique is not determined to be "legitimate," the obligation to provide for the patient's best interest overrides the patient's right to direct his or her own treatment.

Similarly, some argue that a patient's request that heroic measures be avoided, that the prolonged suffering be ended, is not an autonomous choice, but a choice made when one is not truly "in one's right mind." Hence, the decision not to respect such a choice, that is, the decision not to acknowledge one's right to choose to die, is overridden by the determination that to do so would not be "in the best interests of the patient."

The questions of what counts as being in a patient's best interests and who is in the best position to make such a determination are both addressed in many of these essays, in particular those in the sections dealing with death and dying (Part III) and the refusal of treatment (Part II).

Importance of Innocence

The notion of innocence is an important moral consideration. We try not to harm the innocent; we feel a need to protect the innocent. An important theological reason for arguing that the manipulative techniques employed during in vitro fertilization are morally wrong hinges on the belief that the embryo or fetus is an innocent human being and that the right to life of an innocent overrides whatever rights are possessed by those who are not innocent. In other words, on such a view it is morally wrong to discard an embryo, defective or not, because it is an innocent.

In similar fashion, many argue for the protection of very young children against "death decisions" made for them by their parents on the grounds that one cannot choose death for an innocent. Courts have appealed to this notion in ruling that Jehovah's Witnesses cannot choose religious martyrdom for their children (by refusing them blood transfusions); society's moral obligation to protect innocent life overrides parental rights and rights of religious freedom.

Many ethicists, without making a direct appeal to the concept of innocence, argue about the "rights of the unborn," arguing, in effect, that since they cannot lobby for their own interests, since they lack autonomous agency, it is our duty to ensure their welfare prior to and following birth. Interestingly, such arguments sometimes conclude, contrary to what is suggested above, that an infant's right not to be grossly deformed or severely retarded overrides considerations against aborting the fetus.

Frequently in many of these arguments, the concept of innocence and that of incompetence get run together. One judged to be incompetent to decide certain things for himself or herself, that is, one lacking autonomous agency, is then regarded as an innocent. So, for example, when discussing the rights of patients in irreversible coma, an indirect appeal may be made to the concept of innocence. The right of the "innocent" patient to escape further invasive indignities may be said to override family member's rights to demand continued treatment for a loved one. Or, on the other hand, the fact that the patient now has the status of an innocent may make it wrong to discontinue treatment—for to discontinue treatment is willfully to bring harm to an innocent.

Although the appeal to innocence is taken to be an important consideration in determining overriding moral obligations, it can be seen that in many cases it is not that clearly distinguishable from the judgment that a particular course of action is or is not in a patient's best interests, nor is it that easily distinguished from the judgment that we have an overriding moral obligation to minimize (if not prevent) harm.

Maximizing Benefits and Minimizing Harm

The distinction between maximizing benefits and minimizing harm is not always such an easy one to draw, yet each is frequently offered as a decisive consideration in determining what counts as an overriding moral obligation. The judgment that one has an obligation in cases of conflict to choose the course of action that maximizes benefits is a judgment that for the most part coincides with an appeal to the principle of utility: Is the amount of good (the utility) to be produced by this course of action sufficiently greater for all involved than the amount of disutility? Is the risk involved, for example, in nontherapeutic experimentation on mentally retarded children overriden by the benefits to generations of children yet to come?

In a well-known case involving experimentation on mentally retarded children at Willowbrook State Hospital in Staten Island, NY, it was argued that gaining a better understanding of hepatitis and developing better methods of immunizing against hepatitis justified subjecting severely retarded children to the hazards of scientific research, which included, in this case, actually injecting infected serum in order to produce hepatitis in the children being used for the research. Being able to use these children as a "control group," it was argued, would maximize the benefits of such research by guaranteeing excellent medical care for the children involved at the same time that the research yielded information that would benefit future victims of this strain of hepatitis, a disease apparently quite common in institutions of this type.

But attempts to justify the research were really of two types. On the one hand, the researcher and his associates wanted to claim that the benefits of such research were sufficient to override prohibitions against unjustly harming or coercing others. On the other hand, they wanted to argue that inasmuch as children institutionalized at Willowbrook were extremely likely to be exposed to the disease anyway, this particular experiment *minimized* the harm done to these children by inducing a *controlled* case of hepatitis rather than attempting to treat a case contracted in the normal course of remaining institutionalized. In other words, the research was said both to maximize the benefits to others and to minimize the harm to the patients — considerations of sufficient moral weight to override whatever rights these children may have had not to be used as means to some end.

Meeting Social Needs

Though not entirely distinct from the consideration of whether an action maximizes benefits or minimizes harm, this particular standard for determining what counts as an overriding moral obligation is not usually couched in the language of benefits and harms. Rather, its justification for being of sufficient moral weight to override conflicting obligations is usually put in terms of protecting others or maintaining certain social ideals.

The cycle of child abuse, for example, is now a well-documented social phenomenon. Abused children become abusing parents, and so on. The goal of reducing the incidence of child abuse and of breaking the cycle of abuse is now viewed by some as a high-priority goal. Given the facts pointing to a cycle of child abuse, it has been argued that the obligation to ensure that children only be born to parents who truly want children overrides whatever obligations we might be said to have to see that a fetus is brought to term.

Similar sorts of reasons are offered as justifications for aborting fetuses determined to be genetically defective through the use of amniocentesis. The birth of a child, known in advance to be genetically defective, may cause undue financial and emotional hardship to its parents and siblings, severe enough to effect disintegration of the family. The social goal of preserving family relationships may be worthy enough to override the rights of a defective fetus.

In addition, there are occasions when providing for the good of a large group in society is said to override certain rights of single individuals. When we institutionalize someone whose behavior is aberrant or antisocial, we do so believing that preventing that individual from victimizing others is an obligation that overrides that individual's right to freedom. Likewise, an individual's right to expect that confidentiality will be assured by the physician is argued as justifiably overriden in cases when that individual reveals something that poses a threat to society. So, for example, if one in confidence were to reveal to the physician that one had hidden an activated bomb in a nearby concert hall, the obligation to protect those about to attend the concert would justify abridging the right to confidentiality of the patient.

Summary

The fact that we can delineate a number of considerations offered with a view toward determining what rights and obligations take priority in cases of conflicts does little to solve the problems. Many of these standards have been found unacceptable for one reason or another by philosophers, theologians, lawyers, and health professionals alike. But they do offer a starting point. It has been seen that the claim that one has a right to something is not really an absolute guarantee of anything. Other considerations, both moral and nonmoral, enter in. But, for each of these authors rights are viewed as intimately connected with providing the principles of moral action that are essential in the provision of health care services.

Suggested Readings

Annas, George J. "The Hospital: A Human Rights Wasteland," *The Civil Liberties Law Review*, Fall 1974, pp. 9-27.

—. *The Rights of Hospital Patients: The Basic ACLU Guide to a Hospital Patient's Rights*. New York: Avon Books, 1975.

Barry, Brian. *The Liberal Theory of Justice*. New York: Oxford Univercity Press, 1974.

Becker, L. C. "Three Types of Rights." *Georgia Law Review* 13 1197 (1979).

Bedau, Hugo A. *Justice and Equality*. Englewood Cliffs, N.J.: Prentice-Hall, 1971.

Brandt, Richard. *Social Justice*. Englewood Cliffs, N.J: Prentice-Hall, 1962.

Dworkin, Ronal, *Talking Rights Seriously*. Cambridge, Mass.: Harvard University Press, 1977.

Feinberg, Joel. *Social Philosophy*. Englewood Cliffs, NJ: Prentice-Hall, 1973.

Fried, Charles. "Equality and Rights in Medical Care." *Hastings Center Report* 6 (February 1976): 29-34.

—."Rights and Health Care—Beyond Equity and Efficiency." *New England Journal of Medicine* 293 (July 31, 1975): 241-245.

Hart, H. L. A. "Are There Any Natural Rights?" In *Political Philosophy*, ed. A. Quinton. London: Oxford University Press, 1967, pp. 53-66.

Hohfeld, W. *Fundamental Legal Conceptions Applied to Judicial Reasoning*. New Haven: Yale University Press, 1914.

Locke, John. *Second Treatise of Government*. New York: Bobbs-Merrill, 1952.

Macklin, Ruth. "Moral Concerns and Appeals to Rights and Duties: Grounding Claims in a Theory of Justice." *Hastings Center Report* 6 (October, 1976), 31-38.

Nozick, Robert. *Anarchy, State, and Utopia*. New York: Basic Books, 1975.

Olafson, Frederick A., ed., *Justice and Social Policy*. Englewood Cliffs, NJ: Prentice-Hall, 1961.

Pilpel, Harriet. "Minors' Rights to Medical Care." *Albany Law Review* 36 (1972): 462-487.

Rawls, John. *A Theory of Justice*. Cambridge: Harvard Univesity Press, 1971.

Sade, Robert M. "Medical Care as a Right: A Refutation." *New England Journal of Medicine* 285 (Dec. 2, 1971): 1288-1292.

Vlastos, Gregory. "Justice and Equality." In *Social Justice*, ed., R. B. Brandt. Englewood Cliffs, NJ: Prentice-Hall, 1962, pp. 31-72.

Chapter 2

Patients and Their Healers

Historical Studies in Health Care

Guenter B. Risse

Introduction

In the health care field a major issue of concern is the widely perceived conflict of rights between professionals and their patients. Heavily debated questions such as—who should decide about specific aspects of the therapeutic encounter?—reflect a growing dissatisfaction with present models of the patient-physician relationship. Voices are calling for a renegotiation of the social contract between both parties. Some of these attempt to set limits to professional autonomy and suggest greater sharing of decision-making power between the sick and their healers.

At the same time, there is worry that reform will swing the pendulum from traditional paternalistic models of patient-physician association to a "consumer" prototype. Healing roles may be diminished to mere executions of decisions made by society or ill persons themselves.[1] To forestall such developments, physicians' rights have been formulated with the aim to preserve professionalism and public service in a rapidly changing socioeconomic context.[2]

27

Calls for changes in the patient-doctor connection come as no surprise to historians acutely aware of its dynamic character. As debates over the moral issues inherent in these relationships gather momentum, such dimensions should be examined historically.[3] Analysis of the external circumstances and central requirements necessary for healing promise clarification of the issues.

To focus on the history of ethical decision-making in medicine thus requires an examination of the changing relationship between doctors, their patients, and the societal setting of their association. All too often in the past, history has merely investigated discoveries and ethical postures consonant with contemporary information. Values taken out of context were unduly emphasized. However, if history has a role to play in the ongoing debates, it is in the presentation and analysis of complexities inherent in therapeutic relationships as they developed in past societies.

Sickness, however defined, inevitably prompts questions regarding its occurrence, the boundaries of human existence, the meaning of life and death. The ultimate causes of illness might be seen in supernatural beings, evil forces, sinful acts, or simply fate—all explanatory schemes designed to make sense of the event. On such a basis those affected by physical and emotional disease are placed into socially acceptable roles to ameliorate their burdens. Loss of control, anxiety, and withdrawal demand prompt attention designed to remove the prevailing climate of uncertainty.[4]

Healing, in turn is a human activity par excellence, one quite distinct from the mutual grooming assistance provided by higher apes or the instinctive care of mothers for their young. Over a period of thousands of years this function has occupied a prominent place in human affairs, a necessary response to sickness resulting from contacts with pathogenic microorganisms or the hazards of social organization.

Seeking help from another human being in the face of dysfunction, uncertainty, and fear creates an unique relationship between sick people and those who claim to heal. Although the relationship has constantly changed throughout history within the context of shifting societal values and activities, the basic moral nature of healing has remained unaffected: one human coming to the aid of another, trying to ameliorate suffering as best as possible.

Caring for the sick necessarily draws healers into making ethical decisions while attempting to reconcile the conflicting needs of specific patients, society, and health professionals. In terms of roles and norms of behavior, the therapeutic relationship displays a number of opposite qualities.[5] Historically, every society has had a *limited* number of healers, whereas illness is a phenomenon of nearly *universal* dimensions. Moreover, patients are only *temporarily* drawn into the world of disability and sickness, but healers usually remain in such a milieu *permanently* during their professional careers. Next, patients consider the sick role an *involuntary* occurrence, while curers are usually drawn *voluntarily* into their roles. Finally, the healing function carries with it a certain *authority* and *activism* usually in sharp contrast to the *subordination* and *passivity* of the diseased.

The purpose of this paper is to sketch briefly five historical patient-healer relationships. In chronological order an ideal preliterate society is examined, displaying characteristics closely resembling those proposed for prehistoric times, followed by a view of physicians and patients in ancient Greece. The third example comes from the late Middle Ages deeply shaped by the tenets of Christianity, while the next one reflects the new secularism of eighteenth-century Enlightenment. Finally, the last case study focuses on the new scientific medicine that first emerged in Germany towards the end of the nineteenth century. Eschewing a more systematic and complete analysis in favor of a disjointed and fragmentary treatment of the subject reflects present gaps in historical knowledge as well as the rather programmatic character of this essay.

II

Healing relationships that may have existed in prehistoric times are based on hypothetical reconstructions using archeological findings and anthropological studies of modern preliterate societies. In spite of the speculative character of such comparisons, it can be said that these groups display behavior most closely linked to the period of human evolution predating the emergence of writing. In prehistoric societies, healing functions were included among other tasks of shamanism, a form of religious belief and activity aimed at protecting people from destructive actions. The inclusion of sickness within supernatural causality brought it into close association with religion and magic. Through a continuing relationship with the spiritual world, the shaman attempted to avoid conflicts with supernatural powers at the same time providing explanations for events affecting the group. The activities went far in confirming and solidifying a distinctive value system within an animistic worldview.

Thus, sickness was generally interpreted in negative terms. A high fever causing delirium could be variously ascribed to an angry god, unhappy ancestor, or jealous neighbor. In many preliterate societies it still may be seen as a vague, general disharmony regardless of the outward symptoms manifested by individual patients. Like today, other cultures probably only recognized general categories of causality such as witchcraft for all sickness. Although lacking clinical precision in the modern sense or even appearing arbitrary to the casual observer, these causal relationships provided solid and pausible explanations for the events. Moreover, sickness could not only be considered a threat to the individual falling ill, but to the entire social group as a manifestation of hostile forces unleashed through magical action.[6]

Thus, the early therapeutical relationship featured on the one hand a powerful priest-magician capable of manipulating cosmic forces within a logical and determinated system. His (and occasionally, her) selection by the social group, coupled with a long apprenticeship in techniques, authenticated the healing function. A desire or obligation to serve his fellow man and come to the rescue of the sick reflected basic ethical components largely inarticulated. In addition, violations of the social code were assumed to produce illness and demand confessions, thus casting the shaman in the role of a moralist.[7]

The sick person, on the other hand, aware of the supernatural implications of the illness, was expected to react strongly to the disability, adopting a sick role that allowed exceptions from other social functions while rallying broad support from immediate relatives and the entire social group. A ripple-effect of fear and anxiety pervading both the sick person and close companions heightened if the illness was considered the result of the breach of taboo — a violation of certain rules or moral codes leading to supernatural punishment.

The gap in social status and power between healer and patient generally conspired against close personal relationships. In fact, today healing ceremonies in preliterate societies at times are still conducted without the sick person even being present. Given the definition of sickness, the curative efforts assumed a collective dimension. They constituted an opportunity to reaffirm cultural and religious values, reinforcing the bonds of kinship, and sanctioning specific modes of behavior.

Whether exorcised, subjected to magical sucking, purification rites, or seclusion, the patient had great faith in the shaman's awesome powers. This firm belief was a powerful tool for effective healing. Quesalid, the Kwakiutl shaman from British Columbia, writes in his autobiography that his first treatment during the apprenticeship period succeeded because the patient "believed strongly in his dream about me," a dream that had prompted relatives to call Quesalid to the patient's bedside.[8] Thus recognition of high status, acceptance of the healer's role to fight the unfavorable circumstances manifested as sickness, and trust in his capacity to control those forces affecting body and mind, are the necessary conditions bringing patients to preliterate shamans and laying the groundwork for a successful cure.

III

Turning to healing activities that took place in classical Greek antiquity, some fundamental changes in the therapeutical relationship are evident. Most importantly, the conditions of health and disease were excluded from a magico-religious framework and considered part of a lawful "natural" world. The awesome questions of ultimate causality in illness were broadly ascribe to "fate"—either accidents or heredity—thus freeing the affected individual from moral responsibility for its occurrence.[9]

With health preservation and recovery deemed feasible within a rational cosmos, ancient Greek physicians build a fairly complex body of medical theory to explain healing based on prevailing philosophies of nature. A collection of such writings, the Corpus Hippocraticum, became the repository of "classical" knowledge, serving as a basis for both education and legitimacy of the rational practitioner.

The role and social position of the ancient Greek physician changed markedly from that of his priest-magician ancestor. Although literate and a free citizen, he was just a secular craftsman, forced to make a living from healing. Moreover, Greek society was not concerned with the regulation of these practitioners who cured without a license nor adhered to any enforceable code of professional standards.[10] No longer consulted about the *why?* of an illness, they merely were expected to offer council concerning *how* to remain healthy or regain this status, primarily through life style advice involving diet and exercise. Prospective Greek patients, on the other hand, although free from responsibility for the ultimate reasons of their disability, were enjoined to take charge of their health and maintain it through a judicious life style.[11] Their basic knowledge regarding such matters was not very different from that espoused by the self-appointed healers, both believing in a homeostatic system of bodily humors and qualities subjected to environmental influences and diet.

Given the healer's lack of religious sanction and society's disinterest in legislating medical activities, the Greek physician's preoccupation with reputation and fees profoundly shaped the doctor-patient relationship. Since the gap of knowledge between the professional and layman was rather narrow, physicians stressed their clinical proficiency and ability to understand nature. "Love of art" or *philotechnia* was their goal, the vehicle to higher fees and greater reputation. It is precisely the capacity to accurately prognosticate that Greek representatives of the *techne iatrike*, or healing craft, boasted to attract distressed patients.

At the same time, the monetary retribution acquired significant dimensions, prompting the author of one Hippocratic treatise, *Precepts*, to write that "one must not be anxious about fixing a fee, for I consider such a worry to be harmful to a troubled patient, particularly if the disease be acute. For the quickness of the disease, offering no opportunity for turning back (to haggle over fees) spurs on the good physician not to seek his profit but rather to lay hold on reputation."[12]

The same text also struggled to establish a moral basis for the therapeutic relationship. Physicians were admonished: "Sometimes give your services for nothing, calling to mind a previous benefaction or present reputation. And if there be an opportunity of serving one who is a stranger in financial straits, give full assistance to all such. For where there is love of man, there is also love of the art."[13] Such "love of man" or *philanthropia* has been seen as the *a priori* basis of the Greek physician-patient relationship, an unjustified assumption nowhere reflected in the extant medical sources.[14]

At this point it must be observed that the so-called *Hippocratic Oath* does not help in eliciting the ethical foundations of the typical Greek doctor-patient relationship. A document considered to reflect the acme of Greek medical wisdom, it has been extensively quoted, evaluated, and revisited often as a normative guideline for medical behavior.[15] Yet, careful studies have confirmed the basic esoterism of the *Oath* as a series of precepts composed for an exclusive sect of healers, possibly Pythagorean in their philosophical outlook. Rules about surgery, poisons, and abortion were outside the mainstreams of classical Greek medical practice and only acquired greater interest after the advent of Christianity.[16]

What then, were the basic rules governing the classical therapeutical relationship? Physicians felt the need to state that they would help or at least never harm either friends nor enemies. At times public suspicion and prejudice probably curtailed drug dispensing and surgery to reinforce such a position and avoid the label of unpunished killer.[17] Yet, in spite of its uncertainties, the encounter of healer and patient could only take place in an atmosphere of trust and confidentiality. Wrote the unknown author of another Hippocratic treatise: "The intimacy also between physician and patient is close. Patients in fact put themselves into the hands of their physicians."[18] The same need for reassurance in the face of fear and anxiety characterized the Greek therapeutic transaction, as exemplified by the earlier quoted prescriptive document: "a physician, while skillfully treating the patient, does not refrain from exhortations not to worry in mind in the eagerness to reach the hour of recovery .... for left by themselves patients sink through their painful condition, give up the struggle and depart this life. But he who has taken the sick man in hand, if he display the discoveries of the art, will sweep away the present depression or the distrust of the moment."[19]

Thus the classical world forged its own patient-healer relationship on the basis of social equality and lack of religious influences within a naturalistic context. Fee for service, and stress on prognosis were central features. Trust was achieved through reputation, proper etiquette, therapeutic restraint. Obligations to prolong the patient's life were absent.[20] One Hippocratic text, *Decorum*, put it this way: "Medicine possesses all the qualities that make for wisdom. It has unselfishness, humility, modesty, reserve, sound opinion, judgment, discretion, purity, magisterial speech, knowledge of the things good and necessary for life."[21]

IV

With the advent of Christianity the therapeutic relationship sustained fundamental changes both in character and scope. An examination of the late medieval period exemplifies the novel settings, roles and responsibilities. Although physical sickness remained a natural imbalance of bodily humors and qualities, its ultimate cause could be ascribed to divine punishment for the commission of sins. This Judeo-Christian tradition stressed the importance of the patient's moral state—which was accountable for the illness—over that of the physical condition, merely a secondary expression of the presumed transgression.[22]

Therefore, following the 4th Lateran Council of 1215, Pope Innocence III decreed that under the threat of excommunication medieval physicians insist their patients confess to a priest before undergoing medical treatment. In Spain physicians were even fined if they paid two visits to severely ill persons without persuading them to confess. Wrote Arnald of Villanova (1235–1311) of Montpellier: "Physician! When you shall be called to a sick man, in the name of God seek the assistance of the Angel who has attended the action of the mind and from inside shall attend departures of the body . . . . Therefore, if you come to a house, inquire before you go the sick whether he has confessed, and if he has not, he should confess immediately or promise you that he will confess immediately, and this must not be neglected because many illnesses originate on account of sin and are cured by the Supreme Physician after having been purified from squalor by the tears of contrition."[23]

Identified as healers of the body, medical physicians were supposed to approach their task as good Christians. Philanthropia was replaced by love and compassion. Medical intervention was a charitable deed promising its practitioners their own salvation. These values are embodied in a modified version of the Hippocratic oath that begun with the phrase "Blessed be God the Father of our Lord Jesus Christ" and ended with "God be my helper in my life and art." It repeated previous prohibitions on giving poisons, or inducing abortions, while stressing the need for confidentiality.[24] Moreover, early in the 13th century, the University of Montpellier established its own oath, in which social responsibilities, absent from classical writings, were embodied such as free services to the needy.[25]

In marked contrast with conditions prevailing during antiquity, medieval healers in the 14th and 15th centuries were organized into guilds of physicians, surgeons, and apothecaries that represented early stages of professionalization. The restriction of activities and establishment of standards for competency were monopolistic efforts aimed at greater status and remuneration. Rivalries between these organizations and efforts by physicians to control the apothecaries reflected the struggle for greater power and advance of self-interest characteristic of these guilds. During the second half of the thirteenth century, medicine had become an academic discipline in its own right at Bologna.[26] These studies leading to a doctorate and a license to practice further enhanced the professional status of the physician.

Hence, medical practitioners increasingly came under strict regulations established by local governments in conjunction with the various guilds. The purpose was the elimination of irregular healers and quacks who had no medical training. In London, for example, legislation concerning malpractice was initiated by the medical and surgical guilds in the fourteenth century to indicate the legal practi-

tioner's concern for patients' rights while displaying their achievement of certain standards of technical proficiency. The new
accountability lead to a number of malpractice suits dealing with
unjust fees—payments were often promised upon condition of a
"cure" rather than just medical attention, at times even specifying the
time period necessary for the recovery.[27] Other lawsuits were brought
against practitioners demanding compensation for losses or injuries
sustained during treatment. In the case of surgery, such legal actions
even prompted the establishment of malpractice insurance schemes.[28]

Finally, the Christian qualities of the medieval healer were
severely tested as fear of contagion, panic, utter helplessness, and
frustration emerged during the severe epidemics beginning with the
Black Death of 1349. Refusal to visit patients, desertion, and outright
flight from the affected cities were frequently reported.[29] Yet, many
physicians remained and tried to help the sick, especially by writing
tractates in the vernacular with advice on prophylaxis and suggestions for domestic management. Wrote Guy de Chauliac (1300–1368),
a prominent French surgeon: "It was useless and shameful for the
doctors the more so as they dared not visit the sick for fear of being
infected. And when they did visit them, they did hardly anything for
them and were paid nothing. And I, to avoid infamy, dared not
absent myself but with continual fear preserved myself as best I
could."[30]

Here Chauliac was expressing a common dilemma: to treat or
not to treat. Following the first option in the face of an incurable
disease could have been interpreted as a disreputable and greedy
gesture, since such medical problems usually exempted physicians
from any ethical obligation.[31] Refusing treatment, on the other hand,
violated basic tenets of compassion or at times contractual arrangements that individual cities had made with so-called "plague
doctors."

In short, the medieval patient-physician relationship began to
exhibit a number of new features. Although healers claimed new
standards of technical proficiency and a coherent body of theory
ratified by an university degree and a license to practice medicine,
laypersons sought to protect their greater dependency on professionals. Strong doses of scepticism, establishment of strict contractual
arrangements, and legal recourse ensued. In response, healers
expressed a variety of ethical values designed to generate patient
confidence. Compassion, honesty, dedication, lack of greed, and confidentiality were stressed in oaths and declarations, based on the
fundamentals of Christian morality.

V

In Great Britain, the impact of both the population explosion and early industrial revolution of the second half of the eighteenth century created a new awareness about the importance of health in national economic development. Profound social and economic changes facilitated greater demands for health care, threatening the traditional relationship between physicians and their small number of aristocratic clients.[32] As a result, the formally successful division of health care providers along lines established by the medieval guilds —physicians, surgeons, and apothecaries—began to crumble and coalesce into a new role: the general practitioner.[33]

Adding to the confusion of medical roles was a growing uncertainty about the relevance and accuracy of contemporary theories of health and disease, formulated to replace the outmoded classical schemes of humors and qualities. In Britain, the hydrodynamic formulations of the Dutch physician Herman Boerhaave (1668–1738) had given way to neurophysiological schemes offered by William Cullen (1710–1790) and John Brown (1735–1788), both of Edinburgh.[34] Highly speculative in character, these medical systems had little impact on therapeutics, but caused considerable discussions and controversies in professional circles.

With acrimonius intraprofessional squabbles vented in public over theoretical matters, educational requirements, questions of roles and status, the patient-physician relationship suffered. Appeals to Christian values and the Hippocratic oath had lost strength. Given the contemporary perplexities, physicians were moved to signal their competence and honor to the public through the formulation of new codes of professional ethics. At stake was the maintenance or restoration of confidence in the healing abilities of a rapidly expanding medical profession, its integrity questioned in the light of the chaotic conditions.[35]

Among the most prominent figures in the formulation of professional ideals was John Gregory (1724–1773) professor of the "practice of physic" at the University of Edinburgh, then the most famous medical institution in Europe. Gregory sought to endow the patient-physician relationship with novel philosophical underpinnings, freeing it from previous cultural and religious influences no longer meaningful to the secularism and rationality of eighteenth century Enlightenment.

Based on lectures on the subject delivered to his students, Gregory published in 1770 a small book entitled *Observations on the*

Duties and Offices of the Physician. "I shall endeavor," he wrote, " to set this matter in such a light as may shew that the system of conduct in a physician which tends most to the advancement of his art, is such as will most effectually maintain the true dignity and honor of the profession."[36]

Considering the moral qualifications of future physicians, Gregory stressed the human value of sympathy. "Sympathy produces an anxious attention to a thousand little circumstances that may tend to relieve the patient," he declared, and "sympathy naturally engages the affection and confidence of a patient which in many cases is of the utmost consequence in his recovery."[37] Patient and physician should be friends. Sympathy was seen as a property that caused people to identify with the moral sentiments of others.[38] Gregory, thus established a human quality, sympathy, at the core of the therapeutic relationship to guide both healer and the sick. The imperative to cure illness arising from the context of medical practice became a moral good, not just a convenient cultural or economically advantageous tactic.

Gregory's ideas were influential on a whole generation of students who had flocked to Edinburgh from the Continent and America. His son, James Gregory (1753–1821) author of *Memorial to the Managers of the Royal Infirmary of Edinburgh* (1800) and another physician, Thomas Percival (1740–1804) amplified and extended these notions, especially with a view towards a newly emerging arena of frequent patient-physician interaction: the hospital. Guidelines previously recommended to medical students now became regulations binding physicians to certain modes of institutional conduct.

Percival's *Medical Ethics* or "Code of Institutes and Precepts, adapted to the professional conduct of physicians and surgeons," was published in 1803, more than a decade after the author had been requested to produce such a document by the trustees of the Manchester Infirmary. The work dealt with a broad spectrum of issues including the physician's qualifications as a healer, relationship with colleagues, responsibilities to the public, and, of interest here, transactions with individual patients.[39]

Physicians were urged by Percival strictly to observe "secrecy and delicacy" in private and hospital practice. "The feelings and emotions of the patients, under critical circumstances, require to be known and to be attended to, no less than the symptoms of their diseases."[40] Gloomy prognostications were to be avoided if they negatively affected medical management or falsely glorified the importance of a physician's intervention. Moral responsibilities were not to be skirted: "The opportunities which a physician not unfrequently enjoys, of promoting and strengthening the good resolutions of his patients, suffering under the consequences of vicious conduct," wrote Percival, "ought never to be neglected. And his councils, or even remonstrances, will give satisfaction, not disgust, if they be conducted with politeness and evince a genuine love of virtue, accompanied by a sincere interest in the welfare of the person to whom they are addressed."[41]

Hence, the therapeutic relationship within the development of 18th century medical ethics reflected the growing shift towards secularization and the need to codify desirable human values of compassion, confidentiality, and understanding. Confidence in healers surrounded by theoretical uncertainties and internal professional dissension had to be buttressed. With the emergence of institutional depersonalization and experimentation, both privacy and confidentiality were threatened, creating in patients new doubts and apprehensions. The new fiduciary relationship was precarious.

VI

In the closing decades of the 19th century, Germany was wisely acknowledged as the world leader in medicine. The new scientific healing stressed objectivity in its collection of data, discarding the biographical anamnesis together with the patient's personality in favor of specific measurements of biological activity. Robert Volz, a representative figure in German medical circles of the period, aptly wrote in 1870: "Medicine used to be subjective and the physician operated more on account of his personality than his science. Everything was connected with his character. He enjoyed a trust prompted by the personal impressions he caused. The physician of yesteryear came close to the patient. In order to investigate the causes, he had to be psychologist since he dealt with a sick person, not a disease. Now things are quite different. Medicine is after facts, it has become objective. It does not matter who is at the bedside. The physician faces an object which he investigates, percusses, auscultates, and watches while family relationships at his right and left do not change anything: the sick person has become a thing. Since everybody must understand this, all "Hippocratics" disappear. Naturally this approach removes the "Gemütlichkeit" of the healer's housevisits, personal relationships loosen, and physicians are being changed and selected according to the nature of the disease."[42]

In 1896, Julius L. Pagel, noted practitioner who also taught medical history at the University of Berlin, published a series of articles under the title of *medical deontology*. Although they largely reflect conditions in urban settings, Pagel's observations were, nonetheless, characteristic of an era in which the therapeutic relationship underwent drastic changes.[43]

To begin with, Pagel declared that physicians were no longer viewed as real friends of their patients. The role of confidant and life style adviser for the whole family had disappeared by mutual agreement. Patients were disease material, objects for experimentation to further scientific knowledge. In turn, they simply demanded improvement of specific troubles selecting physicians on the basis of the presumed bodily ailments, terminating the association, often rudely, by dismissal or replacement if professional expertise or performance were in doubt.

According to Pagel, medical men's feelings were being repressed, virtually anesthetized under the influence of the new objectivity. Altruism and unselfishness were no longer considered necessary qualities for healers, technical proficiency was. Jacob Wolff, another prominent German practitioner, advised beginning physicians in his 1896 handbook to display their clinical superiority to the patients by constantly interrupting them during the brief history taking session. Upon mention of a certain symptom by the patient, the doctor was urged to rattle off other symptoms likely to be associated with the presenting complaints in order to impress the patient with a display of clinical proficiency.[44] In fact, some physicians proposed the complete abolition of the clinical history except for having patients answer a few precise questions, since personal accounts of illness within psychological and sociological contexts would only detract from the healer's focus on disease.

Two factors were singled out as being especially responsible for the problems affecting the late 19th century patient-physician relationship in Germany. In the first place, scientific medicine had assumed a much more technical character, employing a variety of devices for diagnostic purposes with the result that a growing number of medical auxiliaries diluted the association. The problem was compounded by the proliferation of medical specialists who further narrowed the scope of their interest and competence, fulfilling scientific requirements rather than helping their patients' broader visions of distress.

More importantly, perhaps, was the public's better knowledge of health matters—for Pagel some had become autodidacts in hygiene

—thanks to the wide availability of popular medical writings. Better educated laypersons were more demanding, less personal, and quite skeptical towards the usefulness of health care.[45] Many supported a return to "natural therapy," confused by the excessive claims of new-fangled scientific panaceas such as electrotherapy. "Ninety percent of the population has no understanding concerning the tasks of physicians," wrote one frustrated German practitioner.[46]

If patients were suspicious or outright hostile of their healers, physicians did little to improve their image. Following the new trade ordinances proclaimed by the North German Union in 1869, practitioners were no longer liable to prosecution if they failed to attend patients on demand. The elimination of such legal compulsions to provide medical care was coupled with exemption from business taxes and other obligations.[47] The reforms strengthened the image of the healer-entrepreneur often refusing treatment unless paid in advance, and deciding to settle down on the basis of greater financial potential instead of medical need.

In fact, extolling the advantages of city practice, even Pagel advised practitioners to ration their curative efforts allowing more time for the joys of family life and participation in cultural affairs. Charity begins at home, he wrote, using the untranslatable English expression. Medical self-sacrifices were generally pointless. Physicians must set limited office hours and avoid night calls without violating basic humanitarian principles. Such restrictive practices would not jeopardize incomes, he assured his colleagues.[48]

New ethical dilemmas further complicated the therapeutic relationship. Passage of the 1883 Sickness Insurance Act and the 1884 Industrial Accident Insurance Act pitted patients against healers as the discovery and reporting of malingerers, evaluation of physical disabilities, and demand for certificates of excuse presented conflicts of interest. Moreover, pressures from insurance funds to increase patient loads and prescribe cheaper drugs threatened medical standards of health care, and if not resisted, eroded public confidence.

During the last years of the 19th century, German philosophers such as Max Dessoir (1867–1947) and Friedrich Paulsen (1846–1908), physicians such as Hugo W. von Ziemssen 1829–1902) and Martin Mendelsohn (1860–1930) critically analyzed the ethics of the contemporary patient-physician relationship. Dessoir recognized that the conflicting demands and duties imposed on physicians occurred as a result of conditions arising from the social order. Practitioners had markedly improved their social and legal situation while patients too had gained further legitimacy for the sick role as greater awareness of the complexities of health and disease emerged. Perhaps, wrote another practitioner, the time has come to establish specific rights and responsibilities for both healers and sick people.[50]

Everyone agreed that relations between curers and their clients had markedly cooled. The former approached the subject of their toils like strangers interested in organ pathology, but often oblivious to the real fears and hopes of their patients. No wonder the latter responded with cool detachment and considerable scepticism. The solution was simple: restoration of confidence in medicine. Among the remedies for the "medical misère" German physician stressed a return to the "art" of medicine without surrendering its scientific underpinnings. By this practitioners meant once more the kind of individual and holistic treatment of yesteryear. Moreover, medicine was to clearly signal its limitations avoiding excessive claims and refraining from overoptimistic predictions. Yes, we need a reform, wrote Martin Mendelsohn, "not of scientific medicine, nor skill or fees, but the medical art."[51]

VII

What are, if any, the lessons of history? Can any conclusions be drawn from the various examples? If anything, good history reveals the dynamic nature of the patient-healer relationship. Glimpses at past therapeutic relationships discern their variability and complexity, point out profound differences and striking similarities. Socioeconomic and cultural factors, the state of medical knowledge, the availability of technology, have continuously shaped the role of healers and patients within multiple therapeutic settings.

Yet, at the core of such fluctuating relationships remains the phenomenon of sickness, prompting sufferers to seek help. As seen in the foregoing examples, concepts of disease, healing and sick roles, institutions, all changed markedly throughout history, but the age-old quest remained essentially the same: search for help, reassurance, explanations, care, in order to regain control and achieve social reintegration.

Therefore, guidelines for ethical analysis must be found within the therapeutic relationship as well as the external context shaping it in every society. The contemporary approach, however, has been inadequate since it views the patient-healer relationship as only shaped by values and claims existing outside of it.[52] Perhaps sociological conflict models of doctor-patient interaction, as exemplified in the works of Freidson and Zola,[53] have significantly influenced this tendency, one that is further accentuated by contemporary attitudes towards the medical profession.

If the social contract needs to be renegotiated, history provides many examples of similar reforms carried out under the impact of changing societal conditions. Whatever the recasting of roles might be, planners and advocates must carefully ponder the essential requirements needed for healing to take place. Any discussion of "who should decide questions" in health care would be irrelevant without taking into consideration the universal needs of those cast in the sick role.

References

1. See a recent summary of the issues: Swazey, Judith P., *Health Professionals, and the Public: Toward a New Contract?*, annual oration, Philadelphia, Society for Health and Human Values, 1979.
2. Jonsen, Albert R. *The Rights of Physicians: a Philosophical Essay*, Washington, DC, National Academy of Sciences, 1978.
3. This has been recognized by L. B. McCullough in the article "Historical perspectives on the ethical dimensions of the patient–physician relationship: the medical ethics of Dr. John Gregory," *Ethics in Science and Medicine* **5** (1978): 47–53.
4. The effects of sickness have been lucidly described in Eric J. Cassel's book *The Healer's Art: A New Approach to the Doctor–Patient Relationship*, Philadelphia, Lippincott, 1976.
5. For further details consult: Foster, George M., and Anderson, Barbara Gallatin. "Some Characteristics of doctor and patient roles," in *Medical Anthropology*, New York, Wiley, 1978, pp. 103–104.
6. A useful summary can be found in Ackerknecht, E. H. "Typical aspects of primitive medicine," in *Medicine and Ethnology*, ed. by H. H. Walser and H. M. Koelbing, Baltimore, Johns Hopkins University Press, 1971, pp. 17–29.
7. See Webster, Hutton. *Taboo, a Sociological Study*, Stanford, Stanford University Press, 1942, especially Chapter I, pp. 22–24.
8. Boas, Franz. "Shamanism," in *The Religion of the Kwakiutl Indians*, part II, translations, New York, Columbia University Press, 1930, p. 13.
9. See Temkin, O. "Medicine and the problem of moral responsibility," *Bull. Hist. Med.* **23** (1949): 3.
10. For an overview consult Edelstein, L. "The Hippocratic physician," in *Ancient Medicine*, ed. by O. and C. L. Temkin, Baltimore, Johns Hopkins Univ. Press, 1967, pp. 87–110.
11. An excellent discussion of this subject is Kudlien, F. "The old Greek concept of 'relative health,' " *J. Hist. Beh. Sci.* **9** (1973): pp. 53–59.

12. "Precepts" in *Hippocrates*, with an English transl. by Jones, W. H. S. 4 vols, Cambridge, Mass., Harvard University Press, Vol. I, p. 317. A discussion of fees can be found in Temkin, O. "Medical ethics and honoraria in late antiquity," in *Healing and History*, edited by C. E. Rosenberg, New York, Watson, 1979, pp. 6–26.
13. "Precepts," in *Hippocrates*, Vol. I p. 319.
14. Entralgo, P. Lain. *Doctor and Patient*, transl. from Spanish by F. Partridge, New York, McGraw Hill, 1969, especially pp. 17–23 stresses philanthropia.
15. Bulger, Roger. "Introduction," in *Hippocrates Revisted; a Search for Meaning*, New York, Medcom, 1973, p. 3.
16. Edelstein, Ludwig. *The Hippocratic Oath*; Text, Translation and Interpretation, Baltimore, Johns Hopkins Univ. Press, 1943 (Supplement to the Bulletin of the History of Medicine No. 1).
17. See the discussion presented by Kudlien, F. "Medical ethics and popular ethics in Greece and Rome," *Clio Medica* **5** (1970): 91-121.
18. "The Physician," in *Hippocrates*, Vol. II, P. 313.
19. "Precepts" in *Ibid.*, Vol. I, p. 325.
20. Amundsen, D. W. "The physician's obligation to prolong life: a medical duty without classical roots," *Hastings Ctr. Rep.* **8** (1978): 23-30.
21. "Decorum," in *Hippocrates*, Vol. II, p. 287.
22. See Ell, S. R. "Concepts of disease and the physician in the early Middle Ages," *Janus* **65** (1978): 153-165.
23. Arnald of Villanova, "On the precautions that physicians must observe," in *Ethics in Medicine, Historical Perspectives and Contemporary Concerns*, ed. by S. J. Reiser, A. J. Dyck, and W. J. Curran, Cambridge, Mass., MIT Press, 1977, p. 14.
24. Jones, W. H. S. "From the Oath according to Hippocrates in so far as a Christian may swear it," in *Ibid.*, p. 10.
25. Etzioni M. M. *The Physician's Creed*, Springfield, Ill., Thomas, 1973, pp. 33-34.
26. Bullough, V. L. "Medical Bologna and the development of medical education," *Bull. Hist. Med.* **32** (1958): 201–215. For a more complete view of the professionalization process in medicine consult the same author. *The Development of Medicine as a Profession*, Basel, Karger, 1966.
27. See Cosman, M. P. "Medical fees, fines, and forfeits in medieval England," *Man & Medicine* **1** (1976): 135–144. On the subject of medical fees in medieval Britain consult: Hammond, E. A. "Income of medieval English doctors." *J. Hist. Med.* **15** (1960): 154–169.
28. *Ibid.*, p. 152. See also by the same author "Medieval medical malpractice: the dicta and the dockets," *Bull. NY Acad. Med.* **49** (1973): 22–47.
29. Amundsen, D. W. "Medical deontology and pestilential disease in the late Middle Ages," *J. Hist. Med.* **32** (1977): 403–421.
30. See Campbell, Anna M. *The Black Death and Men of Learning*, New York, Columbia University Press, 1931, p. 3.

31. This situation is explained through quotations from another medieval French surgeon, Henri de Mondeville (1260-1320) in Welborn, M. C. "The long tradition: a study in fourteenth-century medical deontology," in *Medieval and Historiographical Essays in Honor of James Westfall Thompson*, ed. by Cate, J. L., and Anderson, E. N. Chicago, University of Chicago Press, 1938, p. 351.
32. A sociological analysis of the medical patronage system is Jewson, N. D. "Medical knowledge and the patronage system in 18th century England," *Sociology* **8** (1974): 369-385.
33. For details see Hamilton, B. "The medical professions in the eighteenth century," *Econ. Hist. Rev.* **4** (1951): 141-170, and Parry, Noel and Jose. "From apothecary to general practitioner: a successful struggle for upward assimilation and occupational closure 1790-1858," in *The Rise of the Medical Profession*, London, Croom, 1976, pp. 104-130.
34. A general panorama of 18th century medical systems is presented in King, L. S. "Theory and practice in 18th century medicine," *Stud. Voltaire 18th Cent.* **153** (1976): 1201-1218. More details on the medical systems of Cullen and Brown can be found in King, Lester S. "Of fevers," in *The Medical World of the Eighteenth Century*, Chicago, Univ. of Chicago Press, 1958, pp. 139-147. A special article on John Brown is Risse, G. B. "The Brownian system of medicine: its theoretical and practical implications," *Clio Medica* **5** (1970): 45-51.
35. Waddington, I. "The development of medical ethics—a sociological analysis," *Med. Hist.* **19** (1975): 36-51.
36. Gregory, John. *Observations on the Duties and Offices of a Physican*; and on the method of prosecuting enquiries in philosophy, London, Strahan and Cadell, 1770, pp. 9-10.
37. *Ibid.*, pp. 18-19.
38. Gregory's "sympathy" was based on the philosopher David Hume's notion that this faculty was something we all possessed to get an impression of another person's character. For detail see McCullough, L. B.," *op.cit.*, pp. 48-50.
39. Consult the historical introduction written by C. R. Burns to a reprint of Percival Thomas. *Medical Ethics*, Huntington, N. Y., Krieger, 1975, pp. XIII-XXVII.
40. *Ibid.*, III, p. 72.
41. *Ibid.*, XXIX, pp. 107-108.
42. Volz, Robert. *Der aerztliche Beruf*, Berlin, Luederitz, 1870, p. 32-33.
43. Pagel, Julius L. *Medicinische Deontologie: Ein Kleiner Katechismus für angehende Praktiker*, Berlin, Coblentz, 1897.
44. Wolff, Jacob. *Der practische Arztund sein Beruf*, Vademecum fuer angehende Practiker, Stuttgart, Enke, 1896.
45. Pagel, Julius L. *op. cit.*, p. 49.
46. Schmidt, Heinrich. *Streiflichter ueber die Stellung des Arztes in der Gegenwart*, Berlin, 1884, p. 32.
47. For more details consult Fischer, Alfons. *Geschichte des deutschen Gesundheitswesens* 2 vols., Berlin, Rothacker, 1933.

48. Pagel, Julius L. *op. cit.*, p. 48.
49. Dessoir, Max. "Arzt und Publikum," *Zukunft* **9** (1984): 511.
50. Vierordt, H. "Arzt und Patient," *Deutsche Revue* **18** 3 (1893): 110.
51. Mendelsohn, Martin. *Aerztliche Kunst und Medizinische Wissenschaft,* 2nd ed., Wiesbaden, Bergmann, 1894, p. 43.
52. See McCullough, L. B. *op. cit.*, p. 53.
53. For example, Friedson, E. "Client control and medical practice," *Amer. J. Sociol.* **65** (1960): 374–382, and Zola, I. K. "Medicine as an institution of social control," in *A Sociology of Medical Practice,* ed. by Cox, C. and Mead, M. London, Collier-Macmillan, 1975, pp. 170-185.

PART I
Limits of Professional Autonomy

Chapter 3

Goals of Medical Care

A Reappraisal

H. Tristram Engelhardt, Jr.

Introduction

In order to reappraise the goals of medical care, one must first address certain basic questions. One would need, for example, to have criteria for the successful discovery of those goals or for the proper ways of creating or inventing the ends of medicine. Beyond that, one will need to know what will count as medicine or health care. Leon Kass, for example, has criticized much of what modern medicine does, not in terms simply of it being immoral or frivolous, but in terms of it not being truly medical.[1] Here again one will need to know whether the nature of medicine is open to discovery, or whether we as societies fashion the health care profession in order to pursue efficiently the goals we either discover or create. This question of who decides bites home to the core. It is not simply the question of who decides how to apply, or when to use, heroic measures, or who should decide upon the use of new reproductive technologies. Rather it involves the more fundamental issues of who sets the boundaries for the health care professions, of who defines what these professions are, and what the goals are that they should be pursuing.

Put in this light, the problem of the limits of professional autonomy brings one at once to portraying the geography of lines among the professions: where should health care begin and educational and political institutions leave off? How does one apportion among the major institutions of a society the various responsibilities for seeking human well-being? And even more radically, who should decide what the senses of well-being are that one would want to give to the care of the various professions? It should be clear that these are issues marked by controversy. As one approaches them, one will be interested in deciding how such controversies can come to resolution. One should bear in mind that one element, or possible mode, of closure of such debates is cognitive, though often the grounds are noncognitive. That is, there will be considerations that the participants in a controversy will recognize, or take to be, reasons of an intellectual kind that will influence the resolution of controversies about the boundaries and goals of medicine. And in addition there will be factors that play the role of influences, causal factors that close debates. In any particular, concrete dispute, both factors are likely to interplay as a rich web of forces.

Thus, in examining conflicts between societal interests and individual interests in health care, one is likely to discover that there is no univocal sense of, or consensus concerning, what are the interests of society that are at stake. As a result, there are likely to be tensions among the views of these interests held by the state, by the organs of the state, by particular social groups, as well as by particular individuals. These controversies in health care are made even more intense by the fact that the health care profession is not one profession, but a number of sibling professions often quarreling over the professional goals and perquisites involved in health care. One will be forced to ask whether there is any cognitive sense to the ways in which we at present draw the geography of the health care professions, distinguishing among physicians, nurses, occupational therapists, podiatrists, psychologists, Christian Science practitioners, physical therapists, and social workers, among many others. Are the lines at all natural? Or do they reflect for the most part accidental accretions of power and divisions of a professional territory? How one answers this set of questions will as well influence how one proceeds to reappraise and revise the goals of medicine, for it will help us to see what medicine is and what its various goals can be.

In addition, one will be pressed hard to the question of the extent to which the goods of human life are open to discovery by us, or the extent to which we must always, at least in part, create them. If, for example, one believes that by appealing to an ideal observer who is fully informed, vividly imaginative, impartial, dispassionate, and possessing full human knowledge and breadth of perspective, one can construct a definitive list of the important goods of human life and their ranking, then one may in that case be optimistic about discovering the goals of medicine. By using some set of special conditions for moral reflection, and procedures for gaining objectivity and impartiality regarding human goods, one could then hope to discover the goals of medical practice. Such approaches might include a hypothetical choice theory that envisages rational contractors, as in the case of John Rawls's *Theory of Justice*.[2] However, the more that one despairs of that possibility and is convinced that not only are concrete views of the good life likely to be extremely diverse, but that there are no general rational grounds for selecting one, the more difficult the issue will be. That is, if one suspects that the human moral universe is at least in part rationally indeterminate, one will hold that rational individuals may choose widely divergent and materially irreconcilable views of the goals of medicine.[3]

The more that one moves to the second conclusion, the more one will see issues of professional autonomy and of decision concerning the goals of medicine to be issues of procedure. That is, one will not be as concerned about what is decided, as about how it is decided. Undoubtedly, one will have one's own views of what rational individuals should choose. However, one will not hold that there will be general rational grounds strong enough morally to justify coercing others to accept those goods as a condition for their being acknowledged as rational or moral beings. In that case, the problem of reconciling society's interests with individual interests in health care will become a problem of safeguarding human freedom in the process of societal and individual creation of the goals of medicine. This approach will be libertarian. Respecting autonomy will be a condition for a just or moral resolution of debates. It may not necessarily be the goal towards which the debates strive. However, it will function more as a limiting condition or side constraint upon such disputes.[4]

In short, to reappraise the goals of medical care, one will have to take a stand with respect to the nature of those goals and the nature of medicine, the extent to which those goals and that nature is invented or discovered, and the extent to which the moral universe is an enterprise of common creation as much as of common discovery.

In this paper, I will argue that within fairly broad limits the goals of medicine and the nature of medicine are created by us and can with good grounds be drawn quite differently by different rational individuals. This will involve indicating the broad and heterogeneous senses of "therapy" presupposed by medicine in responding to the various complaints brought to its attention, many of which cannot be construed simply as diseases. I will hold that a narrowly construed sense of disease is not adequate to account for the various things done by medicine. Further, I will argue that even narrowly drawn disease concepts are value-infected, at least in part. Integral to the weak normativist view of disease that I will forward will be the argument that those values are socially determined and likely to be subject to some variation among societies and individuals. In short, it cannot be successfully maintained that the goal of medicine is simply to eradicate or to control disease, or that medicine discovers diseases as value-free facts that merit attention, or as value-laden facts that are universal in their claim. Nor is there a single sense of health that could move the enterprise of medicine or health care.

The second part of this paper will address medicine as a profession in order to show that here, as well, the lineaments are as much invented as discovered. Thus, far from having an incorrigible view of where the best interests of patients lie, the health care professions reflect, in part, so I will contend, special interests that may or may not coincide with the interests of the general society or of particular patients. My conclusion, therefore, will be to support a much more libertarian and consumer-oriented view of medicine than would be sustainable if one accepted some of the 19th-century views of medicine as a value-free science that discovers the character of human pathology.

Curing Diseases: Not the Primary Goal of Medicine

The current ideology of medicine portrays it as a science and technology focused on the cure or at least the amelioration of disease. This scientific view has medicine discovering the true nature of diseases, coming to understand their pathogeneses and etiologies, and then moving to prevent them and to cure them. Proper obeisance is of course made to the fact that not all diseases can at present be cured. Moreover, it is acknowledged that various ravages of nature leave us with pains and debilities that medicine cannot completely remedy. As a result, it is understood that medicine must care for those who are diseased and disabled, and not focus singlemindedly upon curing diseases. Even with this accent on care, not simply cure, the desideratum remains etiological treatment of pathologically demonstrable diseases. It is always preferable to find a specific cure for a disease, rather than to engage in symptomatic treatment. Symptomatic treatment, which aims simply at vague symptoms and complaints, is viewed with some suspicion.

An excellent defense of this view is provided in a very well-drawn article by Leon Kass, "Regarding the End of Medicine and the Pursuit of Health," in which the goals of medicine are identified with somatic medicine, with somatically based derangements, and somatically based well-beings. Kass wishes to exclude from medicine *in sensu stricto* problems of sagging anatomies, suicides, unwanted childlessness, unwanted pregnancy, marital difficulties, learning difficulties, genetic counseling, drug addiction, laziness, and crime.[5] Which is to say, Kass presupposes a sense of somatic medicine, or real medicine, as an enterprise directed by a discoverable sense, or set of senses of the pathological. One might think here of arguments in this vein by Christopher Boorse and others,[6] who have attempted to define disease in terms of discoverable senses of function and dysfunction.

Kass and others who subscribe to such arguments question how, for example, medicine can properly be involved in artificial insemination from a donor, when the woman being inseminated suffers from no disease, is in no way dysfunctional.[7] At best, one can say that the couple is not a functioning reproductive unit. However, that would not identify a disease or a dysfunction in any somatic sense (at least in the person "treated"), but rather in some social sense.[8] The same would appear to be the case with many if not most abortions on request. In fact, a great deal of the vexations of the worried well, who consume a substantial portion of the time of primary health care physicians, do not appear to be medical in this more rigorous sense.

There is, surely, an attraction to more restricted senses of health and disease, function and dysfunction, that might provide a more restricted notion of medicine's goals. On the one hand, if one could secure a more restricted notion of medicine, one might restrict the obligations that arise out of rights to health care. After all, insofar as one envisages a comprehensive health care service, one will need to draw lines demarcating what will be supported through such a service, and what will be beyond its bounds. If such a restricted notion of medicine could be discovered, one could then appeal to it in limiting the claims that might be made under a right to health care. On the other hand, one could make more efficient use of the technical skills of physicians, who are highly trained technologists and scientists engaged in procedures that often require much less sophisticated training. Thus, one might envisage other health professionals sustaining the burden of caring for the worried well. One might imagine as well a special profession of abortionists to provide services for the millions of women seeking them.

The recent history of concepts of disease suggests additional reasons why narrower concepts of disease and of medical treatment appear quite credible. If one examines 18th century nosologies that developed under the influence of Thomas Sydenham (1624–1689), one finds catalogs of complaints.[9] Instead of the disease entities with which we are familiar, one finds instead a list of signs and symptoms that could be forwarded by the patient or by others. One finds, in fact, a list of chief complaints. Major categories in such nosologies include fevers, pains, fluxes, and inflammations.[10] As Foucault has indicated,[11] this world was transformed by 19th century pathologists and physiologists such as Xavier Bichat[12] (1771–1802), Francois Broussais[13] (1772–1838), Carl Wunderlich[14] (1815–1877), and Rudolf Virchow[15] (1831–1902). Through their work this world of medical complaints was reorganized in terms of a world of underlying pathophysiological lesions, processes, or entities.[16] As a result, fevers were no longer major categories, but rather were reclassified in terms of their underlying mechanisms. Pains no longer were a major category of medical concern, but were dispersed within a new nosological framework that no longer gave saliency to medicine's treatment of pains. Instead, physicians were directed to underlying pathologies.

The result was an important change in the goals of medicine. For individuals such as Carl von Linnaeus[17] (1707–1778), Francois Boissier de Sauvages[18] (1707–1767), and William Cullen[19] (1710–1790), the accent was upon what was bothering patients or bothering others about patients, and how these problems could be ameliorated. What we term the basic sciences were seen as auxiliary towards those ends. Thus, one pursued causal explanations only if such an etiological account could aid the therapy and care of one afflicted, not because it would reveal the *disease itself*. As Foucault suggests, beginning with such works as Giovanni Baptista Morgagni's (1681–1771) *De sedibus et causis morborum per anatomen indagatis*[20] (1761), the reality of diseases became hidden from the clinician.[21] Medicine was grounded in the basic sciences, which revealed the true reality of the clinical complaints. By implication, the basic sciences became ontologically fundamental to medicine—they grounded and certified medicine's goals and activities.

To appreciate this one must realize that the nosological world was recast by these developments. Signs and symptoms that had never been clinically associated were now seen to be caused by a single set of underlying pathological processes, all owing to the influences of the same causal factor.[22] Clinical syndromes that were previously seen as one disease were also resolved into more than one disease.[23] The result was that understanding basic pathophysiological processes became a cardinal goal of medicine. It suggested that one could discover diseases, and therefore certify clinical categories as biologically orthodox, or well-founded, or as failing to have a biological foundation by an appeal to underlying lesions.

It is through these developments that the scientific cast to the modern goals of medicine emerged. Though Foucault has described this in his book, *The Birth of the Clinic*, it involves much more a subordination of the clinic to the new basic sciences. One might think of the traditional drama of the clinical pathological conference in which the pathologist shows the internist the true nature of the late patient's real affliction. The auxiliary sciences of Sauvages and Cullen thus become basic sciences to which one can appeal in certifying which goals of medicine are truly medical, truly based in the reality with which the basic sciences can deal.

Horacio Fabrega Jr., suggests that this is one of the major differences between the Western biomedical taxonomy of diseases and those taxonomies employed by preliterate cultures and others untouched by such paradigms.[24] It surely marks the approach of modern medicine. For us, it is important to get behind the symptoms, to assign them a physiological truth value, and in terms of that truth value erect a hierarchy of symptom authenticity.[25] Thus, for modern scientific medicine, dealing with the worried well is suspect in a way it was not for Sauvages, who dealt with categories such as "wandering pains that do not have a name from a fixed site,"[26] or Cullen, whose nosology listed chronic and acute nostalgia.[27] For such physicians, one finds mixed with "orthodox" medical complaints the various problems of the worried well and the vexed ill.

Attempts to reappraise the goals of medicine must start then with the acknowledgment that we have completed a major recasting of medicine and health care in terms of the basic sciences. We have witnessed the establishment of an, in part, very useful medical ideology based on the preeminence of the basic sciences. However, this view accounts for only a portion of what medicine has traditionally done and what it indeed continues to do. Medicine throughout its history has endeavored to aid the sterile in having children and help the fertile to avoid having children too often or under undesirable circumstances. It has attempted to make the ugly look better, and the vexed feel better. Medicine has been practiced at least partially in the image of Dr. Feelgood, a role Kass criticizes.

This should not be too startling if one reflects upon what would be required to identify any given state of affairs as a disease. To identify a disease, to discover pathology, one must find suffering, as the terms pathology suggests. However, the complaints brought to medicine, the sufferings brought to the attention of physicians, are diverse.[28] Some are complaints about pains or discomforts. These are often brought to the attention of the physician without any appeal to the concept of disease. Thus, teething, childbirth, and the symptoms of menopause are usually "treated" by physicians without holding that a disease is involved.[29] Physicians are also enlisted to correct deformities or disfigurements that are held to be vexatious departures from some standard or standards of human form or grace. These complaints range from those of supernumerary fingers, sagging facial features, and strangely shaped noses to those of vitiligo. In short, a great number of the complaints that medicine addresses turn on pains and discomforts, and on failures to achieve certain esthetic norms for the body.

In addition, complaints cluster around various failures to be able to adapt or to function in ways that people expect. However, if one holds an evolutionary view of function and dysfunction, such failures will always be understood within a context. That is, function, successful adaptation, can be understood only within a particular environment that provides the conditions within which physiological and psychological processes transpire. Moreover, there are no standard environments for humans. In addition, one will need to specify the purposes one holds that the capacity or trait should serve. If one is interested in maximizing reproductive fitness, one will see traits and capacities as being functional or dysfunctional, depending on whether or not they maximize reproductive advantage. If one is concerned about maximizing inclusive reproductive fitness, one will be willing to tolerate one individual's failure to reproduce if that failure is tied to the successful reproductive endeavors of his or her relatives. But such is the case only if one is interested in a gene surviving or a species enduring. One may in fact have no such interests, or not rank such interests very highly. Which is to say that one may decide that certain human traits or capacities are functional in the sense of supporting goods that a group has interest in, even if they do not maximize the chances of the species surviving. As a consequence, the goals of medicine cannot be specified simply in terms of discoverable functions or dysfunctions. Medicine is brought to serve individual inclinations or the inclinations of a particular group. It will of course be the case that a great number of biological happenings will undermine so many possible human goods that most individuals and most societies are likely to see them as diseases.[30] However, the point remains that certain processes are perceived as pathological because they underlie states of affairs that we disvalue[31] (and that we also take not to be within the immediate volitional control of the person suffering from the state of affairs), and that we in addition take to be embedded in underlying physiological or psychological processes; i.e., we distinguish diseases from malingering and possession by the devil.

These considerations suggest that the goals of medicine are created by individuals and groups through the complaints they make to physicians. In a sense, conditions are defined as medical conditions if medicine can alleviate the complaints in question. Medicine thus represents a collage of various strategies for coming to terms with a collection of complaints. This collection has no single unifying goal. The most that one could say is that medicine deals with complaints that are not immediately under voluntary control, and the appearance of which is accounted for in terms of pathophysiological and pathopsychological generalizations. Medicine is thus best viewed as a set of strategies for dealing with human problems and for establishing human well-beings. But again, there is not one sense of well-being that is at stake. Instead, there are various senses of freedom from pain, from vexation, and of establishing reasonably attractive human bodily form, or of preserving or of restoring the capacity to achieve the things that individuals or social groups take to be proper to the life of humans. This collage of strategies, which is medicine, will thus have fuzzy boundaries with other major social institutions such as education and politics. As a consequence, the hope of univocally displaying the end of medicine is, *pace* Leon Kass, futile.

Instead, one will specify the goals of medicine by creating lines separating education, medicine, and the law. Where the lines will be drawn will depend upon various concerns about abuses of power within particular institutions and upon judgments about which institutions are likely to be most efficacious in achieving our goals of well-being. It will also depend on the ways in which we assign importance to possible human goals. Reappraising the goals of medicine will thus force us to reappraise the ways in which we rank the importance of human goals. If we wish to maximize the pleasures of this brief life we may, for example, instead of investing in expensive acute and chronic care, invest those funds in the provision of free Bacchanalian revels, the composition of excellent symphonies, and invest only in, or for the most part only in, preventive health care. On the other hand, one might be inspired by the myth of the six million dollar man, woman, and dog, and decide to pursue the extension of human life as far as possible at the cost of foregoing the pleasures of good Scotch whiskey, the virtues of philosophy, and the opportunity to take summer vacations. One will then have created substantial expectations with regard to what will count as ordinary health care.

Thus, if one is confronted with a decision between one's society providing preventive medicine, abortion on request, cosmetic surgery, but only very limited acute and chronic care (giving instead the funds to the pleasures of this all-too-brief life), versus investing those funds in substantial research and care of acute and chronic diseases, there will be no way simply to discover an answer. These goals are as much created as discovered. Such choices in part will turn upon basic philosophical theories about entitlements to goods and the moral significance or insignificance of the natural lottery. However, the goals of medicine will not be usefully sought as hidden essences, nor will a report of how they are viewed by certain communities resolve disputes concerning their "true" nature. Instead, one will be stipulating or selecting certain goals to support major choices concerning the nature of the good life.

Medicine as a Profession

The heterogeneous nature of medicine's goals should suggest that there is no clear unity to the profession or professions of medicine. If it is the case that professions, or at least the learned professions, develop around a complex nexus of special abilities, skills, and needs, then, unless there are natural geographies of needs to order those skills, their actual lineaments will owe more to historical forces than to conceptual necessity. Thus, if one considers the four traditional professions of the law, theology, medicine, and the military, one will see that each offers a disparate set of services.

Some lawyers are advocates of defendants in criminal cases, others assist individuals in seeking recompense for damages, yet others function simply as general counselors in business and other enterprises. The profession of theology encompasses individuals not only concerned primarily with basic conceptual issues, but also those who mediate with the Deity and administer ecclesiastical structures. Actual professions, as one finds them in vivo, usually proffer a broad set of different services. This has been argued above with respect to medicine. Medicine provides care and cure for various complaints of pain, deformity, and dysfunction. It offers various ways of achieving a well-being of freedom from pain, of proper human form, of the accomplishment of functions deemed appropriate for a particular sex at a particular age.

In addition, medicine as a profession is developed around a series of other special professional goals. To begin with, professions are ways in which individuals make a livelihood, sustain themselves, and assume a position in a society. Medicine is, when one considers its inwardly directed goals, a way for its members to acquire wealth, social status, and social power. As a learned profession, it offers, in addition, special intellectual joys. Which is to say, as knowledge is the joy of the knower and art the joy of the artist, so also it is a pleasure to perform the skills of medicine in and for themselves. This point is put in parody when one thinks of a surgeon stating that the operation was a success though the patient died. Yet such an attitude within bounds is one proper to all learned professions. They encourage a pride in the development for its own sake of an intellectually directed skill.

The extent to which a physician, nurse, or other health professional sees a particular request for services as requiring a particular skill will influence the acceptability of that request by the professional. All else being equal, a professional who seeks the intellectual joy of practicing a skilled or learned profession has a ground for rejecting requests that do not require such skill or learning. As a consequence, physicians and other health professionals may not be sympathetic to requests by the worried well for treatment or care when there is nothing that "can be done" through appealing to their special expertise and learning as professionals. Thus, one has another point at which there are likely to be tensions between the expectations or desires of patients and those of health care practitioners. Patients may have different goals in mind for the health professionals than the members of those professions have. Patients may, for example, still want to have physicians and other health personnel play a priestly role, or a more generally caring and supportive role, roles which such professionals were more willing to bear in the past when their skills did not so well establish them as special scientists and technologists.

The same may be the case because of different views regarding what substantiates a complaint as a *bona fide* medical complaint. Since physicians can function as expert cosmetic surgeons, expert abortionists, sacerdotal counselors of the worried well, and arrangers of adoptions and of artificial insemination from donors, patients may want the profession of medicine to embrace the goals of those procedures among its goals. Medicine, on the other hand, may wish to view itself with a scientific purity that appears in the image and likeness of an art based in value-free basic sciences. It is unlikely, though, that all physicians would wish to embrace this narrow view of medicine's charge. Indeed, it would mark a departure for medicine's traditional role of caring for the worried well and using its technological abilities to help people feel better and to have their bodies serve their goals better. Medicine has, after all, traditionally provided abortions and contraception and suggested ways in which infertility can be cured.[32] In fact, some of the very physician-scientists who played a major role in the development of the basic sciences endorsed as well expansive roles for medicine in assuring human progress and in bettering human life.[33]

As a result, an attempt to define the goals of medicine will have to take account of the various ways in which the medical or health care professions view their own and society's needs, as well as the various ways in which society can see the medical professions as being of possible use for society's goals. There is no reason to anticipate unanimity on these matters. In fact, it is most likely that both individual physicians and patients will see the goals of the professions quite differently.

In this regard, it is worth noting that many of the sibling rivalries among the various medical professions (think here of the contentions over territorial prerogatives among physicians and nurses) are much more grounded in historical happenstance and professional interest in status, prestige, and income, than they are in basic conceptual differences. Or to take another example, one might reflect on the fact that psychoanalysts are usually trained as physicians, though they make very little use of their somatic training, while dentists, who share a great deal of skills and needs for knowledge with physician oral surgeons, are trained in separate schools. In short, one should not regard the distributions of goals, duties, and perquisites among the various health professions as simply reflecting basic conceptual realities, but rather as in great measure the end product of various historical and economic forces.

Changing Professional Roles and Changing Goals

This sketch suggests that the goals of the various health professions, and their perquisites and prerogatives, have developed out of varying perceptions of the interests of the health care professions, the interests of society, and the interests of individuals. On the one hand, if medicine and the other health care professions give a special accent to the importance of practicing learned skills, there will be sympathy for the view that health professions have a privileged capacity to determine and choose what is in the best interests of patients. Professionals will have an understandable interest in making good use of their skills and of pursuing the goods to which their most esteemed skills are directed. Unquestionably, health professions have a unique view of the consequences of various complex health care choices. Physicians and nurses are expert counselors of finitude in that they know in painful detail what the consequences will be, or are likely to be, given different choices of therapy. This social distribution of knowledge, a phenomenon described by Alfred Schutz,[34] ineradicably puts the health professional in a privileged position as expert. However, it confers no special privilege in understanding how a patient should weight the consequences of various choices.

The combination of this expertise of knowing the consequences of choices, and the absence of a special privilege of weighting those consequences, has argued for the practice of free and informed consent. The professional should, as far as possible, lay out the geography of consequences and allow the patient to assign values to the various possible outcomes. Unquestionably, the use of free and informed consent erodes the sacerdotal role of the health professional and moves the profession closer to that of the provider of a service. Priests direct those in need to salvation. They are not merely providers of services. The partial eradication of the sacerdotal role of the profession, however, moves health care professionals closer to the role of those who provide services to others. It should, however, be noted that this is not a new phenomenon. It is only that there are more such "requests for service" given a more broadly accepted view

of individuals as free arbiters of their own fates. In addition, such a
view contrasts with an ideology of medicine that would see medicine
as discovering through the basic sciences, in a value-free fashion, the
goals of medicine.

The more that one comes to view patients as free agents, the
more one will be pressed to see the medical professions as ways of
reestablishing patient autonomy over their bodies and minds. In fact,
insofar as one sees the ethical community as one bound together on
the basis not of force, but of respect for the freedom of others, one will
come to envisage the relationship among medicine, society, and
individuals as one of a more or less formal negotiation about goals.
On the one hand, respecting the freedom of patients will commit one
to a form of consumerism. It is in the end the patient who decides
what is worthwhile for him or for her. Medicine is thus available as a
possible means of establishing the patient's autonomy over his or her
body or mind as a way of achieving a patient's view of his or her own
best interests or well-being. Of course, the patient may in the end
wish freely to reject the pursuit of all such goals as in the case of the
competent suicide, or with the refusal of life-saving treatment. On the
other hand, professionals will have their views as to what endeavors
are more proper or more ennobling for their profession. Therefore,
professionals are likely to reject certain requests as menial, as the
kind of service performed only by a technician, or as not really
involving a disease process and therefore not really being medical.
These are the professonal's attempts through his or her own choices
to create a profession that will support his or her view of the good
life, including the "good" practice of that profession.

As I have argued, it is very unlikely that there will be agreement
among patients or among professionals on these matters. It is also
unlikely that there will be agreement between patients and the
professionals. As a result, a society committed to respecting
individuals as free agents will, to begin with, attend with care to the
process of patient-professional negotiation. It will guard rights to free
and informed consent, and therefore acknowledge the correlative
right to refuse life-saving treatment, or for that matter any treatment,
and will as well protect the rights of patients and professionals to
contract for the performance of procedures that many in the society
would find deviant or obscene. One might think here of forbearing
from interfering with individuals pursuing amniocentesis and
abortion for sex selection, artificial insemination of lesbians who
wish to bear children, and the treatment of homosexuals, not in order
to make them heterosexuals, but more effective homosexuals.

Beyond that, such a society will probably value freedom, not simply respect it as a side constraint defining an ethical society. As a consequence, such a society will support institutions and professionals through its common funds and require that, should they participate, they must provide certain services on request. One might think here of the disputes concerning the provision of sterilization and contraception in hospitals built with federal funds, [35] or abortions on request in government-run hospitals.[36]

This view of the professions, individuals, and their societies, gives accent to a negotiation predicated upon (1) the fact that humans have varying views of the good life and of health and well-being, and that there is no univocally justifiable way of establishing the nature of the good life or of the diseases medicine should treat, and (2) the moral obligation to respect the freedom of persons in such negotiations.

Conclusion

A reappraisal of the goals of medicine thus leaves us with the problem of deciding how we as free individuals can shape professions to support the goods and ways of life we take to be important. Since there will be a great disagreement about such choices, we will be forced to establish procedures for tolerating divergent choices. We will need as well to find mechanisms that will prevent particular professions, the state, or elements of society, from coercively dominating those negotiations, while recognizing rights to free association in professions and social groups dedicated to particular goals.

The question of deciding who decides is thus likely to be transformed into two questions: first, deciding how to respect freedom of choice. Here one might think of the Supreme Court rulings acknowledging the right of a woman to seek an abortion.[37] Second, deciding how far society is obliged to support such choices. One might think here of the controversy with regard to the public funding of abortion for the indigent. We are together as free individuals creating the goals of medicine, even though this creation is likely to be marked by contest and dispute. It is the serious endeavor of fashioning our views of the good life. Even where we cannot agree upon the goals, we must attend with care to the process and procedures of respecting free choice in their creation.

References and Notes

1. Kass, Leon. "Regarding the End of Medicine and the Pursuit of Health," *Public Interest* **40** (Summer 1975): 13–14.
2. Rawls, John. *A Theory of Justice*, Harvard University Press, Cambridge, Mass., 1971.
3. It is the view of the author of this paper that this is indeed the case: reasonable individuals, even under ideal circumstances, may disagree concerning the fundamental goods of life. As a result, an ethical community, in the sense of a community based on reason-giving and peaceable manipulations, not force, may be established only through procedures for negotiating moral institutions and views of the nature of the good life, and surely not through procedures for the imposition through force of one view of the good or of the good life upon all the members of the community. Engelhardt, H. T., Jr. "Personal Health Care or Preventive Care," *Soundings* **63** (Fall 1980): 234–256.
4. For a discussion of freedom as a side constraint, see Nozick, Robert. *Anarchy, State and Utopia*, Basic Books, New York, 1971, pp. 30–34.
5. *Op. cit.,* Kass, p. 11.
6. Boorse, Christopher, "Health as a Theoretical Concept," *Philosophy of Science* **44**:4 (December 1977): 542-573. Boorse, Christopher. "On the Distinction Between Disease and Illness," *Philosophy and Public Affairs* **5** (Fall 1975): 49–68; Boorse, Christopher. "What a Theory of Mental Health Should Be," *Journal for the Theory of Social Behavior* **6** (1976): 61–84; Boorse, Christopher. "Wright on Functions," *The Philosophical Review* **85** (January 1976): 70–86; von Wright, Georg Henrik. *The Varieties of Goodness*, New York, The Humanities Press, 1963.
7. Kass, Leon R. "Babies by Means of In Vitro Fertilization: Unethical *Experiments on the Unborn*," *New England Journal of Medicine* **283** (November 18, 1971): 1176–1177.
8. This view often presupposes an unargued assumption that a condition must be a disease in order to justify medical treatment. This, however, is not the case. Medicine often responds to complaints without first legitimizing them as complaints rooted in diseases. For example, physicians now, as for at least two millenia, provide contraceptive advice and materials.
9. Sydenham, Thomas. *Observationes Medicae Circa Morborum A-cutorum Historiam et Curationem*, London, G. Kettilby, 1676.
10. One might consider, for example, Sauvages' classification which listed 10 major classes of diseases: defects (vitia), fevers (febres), inflammations (phlegmasiae), spasms (spasmi), problems with breathing (anhelationes), debilities (debilitates), pains (dolores), madnesses (vesaniae), fluxes (fluxes), and constitutional disorders (cachexiae), appearing in Sauvages de la Croix, Francois Bossier de. *Nosologia methodica sistens morborum classes juxta Sydenhami mentem et botanicorum ordinem, 2* vols., Amsterdam, Fratrum de Tournes, 1768.

11. Foucault, Michel. *Naissance de la Clinique*, Paris, Presses Univer-sitaires de France, 1963. Trans. by Sheridan Smith, A. M. *The Birth of the Clinic: An Archeology of Medical Perception*, New York, Random House, 1973. In his investigation Foucault does not explicit-ly analyze the contributions of Carl Wunderlich and Rudolf Virchow.

12. Bichat, Xavier. *Pathological Anatomy*, Philadelphia, P. Grigg, 1827, p. 14. Trans. by Joseph Togno.

13. Broussais, F. J. V. *De L'Irritation et de la Folie*, Paris, Delaunay, 1828, trans. by Thomas Cooper, *On Irritation and Insanity*, Columbia, South Carolina, McMorris, 1831, pp. ix–xi.

14. Wunderlich, Carl A. "Einleitung," *Archiv fur physiologische Heil-kunde* 1 (1842): ix.

15. Virchow, Rudolf. "One Hundred Years of General Pathology (1895)," in *Disease, Life, and Man*, Stanford, California, Stanford University Press, 1958, p. 192. Trans. L. J. Rather.

16. A great many of these arguments centered around particular nomi-nalist versus realist understanding of the nature of diseases. See Engelhardt, H. Tristram, Jr. "The Concepts of Health and Disease," in *Evaluation and Explanation in the Biomedical Science*, pp.125–143, editors Engelhardt, H.T., and Spicker, Stuart F., Reidel, Boston, 1974.

17. von Linnaeus, Carl. *Genera Morborum*, Upsaliae, Steinert, 1763.

18. *Op. cit.*, Sauvages.

19. Cullen, W. *Synopsis Nosologiae Methodicae*, Edinburgh, William Creech, 1769.

20. Morgagni, Giovanni, *De Sedibus et Causis Morborum* per anatomen indagatis, Venice, 1761.

21. *Op. cit.*, Foucault—see especially chapters 8 and 9.

22. Thus, consumption, or phthisis, and scrofula were seen to be mani-festations of the same disease: tuberculosis.

23. E.g., typhus and typhoid were distinguished as two separate dis-eases rather than being conflated.

24. Fabrega, Horacio, Jr., "Disease Viewed as a Symbolic Category," pp. 79–107. *Mental Health: Philosophical Perspectives*, Engelhardt, H. T., Jr., and Spicker, Stuart, eds., Reidel, Boston, 1978.

25. *Ibid*, pp. 104–105.

26. *Op. cit.*, Sauvages, Vol. I, p. 94.

27. Cullen, William. *Nosologia Methodica*, Edinburgh, S. Carfrae, 1820, p. 121 and 152.

28. Engelhardt, H. Tristram, Jr. "Human Well-Being and Medicine: Some Basic Value Judgments in the Biomedical Sciences," in *Science, Ethics and Medicine*, pp. 120–140, ed. by Engelhardt, H. T., and Callahan, Daniel, Institute of Society, Ethics and the Life Sciences, Hastings-on-Hudson, New York, 1976. Engelhardt, H. Tristram, Jr. "Doctoring the Disease, Treating the Complaint, Helping the Patient: Some of the Works of Hygeia and Panacea," in *Knowing and Valu-ing: The Search for Common Roots*, ed. by Engelhardt, H. T. Jr., and

Callahan, Daniel, Institute of Society, Ethics and the Life Sciences, Hastings-on-Hudson, N. Y., 1979, pp. 225–250.

29. King, Lester. "What is Disease," *Philosophy of Science* 21 (July, 1954): 193–203. Of course, some have indeed argued that conditions such as menopause are diseases. See, for example, Kistner, Robert W. "The Menopause," *Clinical Obstetrics and Gynecology* 16 (December, 1973): 107–129; Wilson, Robert A., Brevetti, Raimondo E., and Wilson, Thelma A. "Specific Procedures for the Elimination of the Menopause," *Western Journal of Surgery, Obstetrics and Gynecology* 71 (May-June, 1963): 110–121.

30. Thus one has a range from great agreement to potential disagreement, from conditions such as sarcoma, which have consequences that will be devalued in nearly all environments, to those such as color blindness, which have advantages in some real environment contexts, and may be valued in such contexts. On this last point, see, for example: Post, Richard H. "Population Differences in Red and Green Color Vision Deficiency: A Review, and a Query on Selection Relaxation," *Eugenics Quarterly* 9 (March, 1962): 131–146.

31. See the discussion of these points in: Margolis, Joseph. " The Concept of Disease," *The Journal of Medicine and Philosophy* 1 (1976): 238–255.

32. "The Physician's Obligation to Prolong Life: A Medical Duty without Classical Roots," by Darrel W. Amundsen, in *Hastings Center Report* 8 (4) (Aug. 1978): 23–30.

33. A good example is Virchow, Rudolf. "Die naturwissenschaftliche Methode und die Standpunkte in der Therapie," *Archiv fur pathologische Anatomie und Physiologie and für klinische Medizin* 2 (1849): 36–37.

34. Schutz, Alfred, and Luckmann, Thomas. *The Structures of the Life-World*, translated and introduced by Zaner, R. M., and Engelhardt, H. T., Jr., Evanston, Northwestern University Press, 1973.

35. Allan Guttmacher Institute and Planned Parenthood Federation of America. Research and Development Division. *Family Planning, Contraception, Voluntary Sterilization and Abortion: An Analysis of Laws and Policies in the United States, Each State and Jurisdiction*, Rockville, Md., Bureau of Community Health Services, US Department of Health, Education, and Welfare, 1978, pp. 3–48.

36. *Ibid.*, pp. 9–15, 79–88. Also: Right to Choose v. Byrne, 398 A. 2d. 587 (N. J. Supr. Ct., Jan. 10, 1979).

37. Roe *v.* Wade, 935, S. Ct. 705 (1973) and Doe *v.* Bolton, 935. S. Ct. 739 (1973).

Chapter 4

Establishing Limits to Professional Autonomy: Whose Responsibility?

A Nursing Perspective

Mila Ann Aroskar

> *It was the best of times, it was the worst of times, it was the season of Light, it was the season of Darkness, it was the spring of hope, it was the winter of despair.*
>
> **Charles Dickens**

The richness and complexity of the kaleidoscopic social phenomenon known as professional nursing requires viewing the nurse and nursing as embedded in a complex network of relationships at all levels of the health care system. This recognition precludes any simple or single answer to the question of responsibility in setting limits to professional autonomy. It seems useful at this time to look at various facets of the question, that is, some of the pieces that one puts into one's kaleidoscope as one looks at the dimensions of autonomy. The asking of any question that involves multiple relationships and interactions risks finding several possible answers that then require significant decisions and choices having far-reaching implications for individuals and groups. Often "solutions" challenge tenaciously held assumptions and values. This question has proved to be no different as one journeys from the past into the present and future.

Historically, Nightingale, the founder of modern nursing, visualized the responsible professional nurse as acquiring specific knowledge and developing character, both essential aspects for carrying out a significant social service. These characteristics are also related to autonomy and accountability and the values one holds in relation to these characteristics. A century ago, nursing was generally viewed as subservient to medical care, a view *not* held by Nightingale. In 1932, Goodrich spoke of the nursing profession as itself intrinsically ethical and an essential factor of the social order. She described the nurse as a public servant and acknowledged that the field of nursing is under the direction of the doctor. These two nursing leaders in different eras reflect the *intra*-disciplinary conflict that still exists related to nursing autonomy. For example, one point of view is reflected in remarks that include nursing under the rubric of medicine and another perspective in remarks about a health care system that includes medicine, nursing, and other disciplines.

The struggle for more autonomy by some in nursing is an on-going challenge as more nurses seek to develop coworker and collegial relationships with physicians and other health workers in bureaucratic structures. Varying definitions of nursing and the paradoxical strands of obedience, consumerism, and professionalism in nursing make it unclear precisely what nurses or the nursing profession should be autonomous about in a strict sense, and for what purpose autonomy is sought. Is it for a self-seeking end, as it sometimes appears, or for some general goal of improved service to clients. Or is it an amalgam of the two that respects both the provider and the client?

Although there is no consensus on one definition of nursing, for purposes of this paper I shall consider nursing to be assisting individuals, families, and communities in situations of health crises or of crisis prevention. This is accomplished through such functions as case finding, assessment, health teaching, individual or family health counselling, physical and psychological care supportive of life and well-being, therapeutic interventions, and restorative services.

Nursing provides resources directly and indirectly for meeting health needs when the client is unable to do so.

In reflecting on the development of more autonomy in nursing and where responsiblity should lie for limiting professional autonomy, a dangerous assumption is sometimes made. The assumption is that anyone labeled "nurse" has the same educational background, skills, and talents as anyone else with that label. Talking to adult nurse practitioners in a clinic, a staff nurse in an ICU, a community health nurse, and a nurse psychotherapist might leave one feeling a bit uncomfortable about this assumption. According to Lambertsen, the registered nurse group can be divided into two subgroups: those who are prepared to function within existing structures or defined patterns of practice and those prepared to function within unstructured or ambiguous patterns of practice, commonly, the nurse practitoner.[1] Degree of risk, dimensions of control affected by either the presence or absence of a physician, and the nature of clinical judgement required are aspects that must be considered in looking at the autonomy of decision-making in these two subgroups. The assumption that registered nurses represent a homogeneous body of similar talents is incorrect and according to Lambertsen impedes understanding of the potential scope of nursing's contribution in the present and emerging health care system: legally, philosophically, and practically.

Regardless of the subgroup to which the individual nurse belongs, most nurses and nursing personnel are employees in complex bureaucratic organizations. They work in an interface position between the patient and another party such as the physician or hospital administrator. The complex interface position of the nurse can be seen further in the multiple sources of accountability that affect autonomy: accountability to self, to client, to physician, to employing agency, to the profession, and to the state. The interface position often creates conflict for nursing between bureaucratic and professional models of practice. Bureaucratic organizations are characterized by an administrative hierarchy concerned primarily with the maintenance of administrative structure and secondarily with achieving organizational goals, e.g., service to clients. Under the professional model, decisions and actions are dictated by the needs and circumstances of clients, which take priority over organizational rules and procedures. If professionals do not have autonomy *and* accountability for professional decisions, administrative authority and accountability will probably take precedence.[2]

Autonomy and Accountability: The Same Coin

Accountability and autonomy are opposite sides of the same coin of professional practice in the sense that one can hardly be accountable if one is simply a handmaid to someone else, i.e., in a professional, legal, and moral sense. Autonomy has to do with the right of self-determination, governance without outside control, and the capability of existing independently. In a more existential sense, the individual is free to make choices and then bears the consequences of those choices. The aspect of independent existence places some immediate limitations on the autonomy of any professional group since health care needed by any one individual or group often requires the technical expertise of several disciplines. However, the concept of professional autonomy requires that practitioners be self-regulating and have control over their own functions in the employment setting. At present, many nurse practitioners are confronted daily with the reality that others are or consider themselves to be in control of nursing until challenged.

Accountability has to do with responsibility and answerability to another for one's choices and actions. The accountable individual is prepared to explain and to take consequences of his or her decisions and actions, again implying the autonomous nature of these decisions and actions. This concept does not support the notion of providers "covering up" for each other when errors have been made. In the professional model, accountability for behavior coexists with increasing autonomy. Both concepts have individual and collective dimensions. Professionals gain and maintain control over practice by demonstrating answerability to their clients, i.e., the public. Ways in which professionals collectively assure safe practice in the best interest of clients include: the development of codes of ethics that provide guidelines for defining professional responsibility in the client/provider relationship; the establishment of rigorous qualifications for entry into practice; the requirements of peer standard setting, e.g., *A.N.A. Standards of Practice*; and the development of legal definitions of liability and practice acts that inform the public of the duties and obligations of professionals.

If nurses are to realize this complex of autonomy and accountability in practice, then organizational structures must be developed in which this can become a reality with the ends in view of more effective nursing service for clients and the integrity of the professional practitioner. Most health care settings, as currently operated, do not foster the development of autonomous and accountable collegial nurse practice. For example, nurses have been delegated administrative authority and accountability, but have not been granted, or taken, responsibility for their own patients in most institutional settings, a catch-22 situation. Traditionally, nurses have not been expected in many settings to make professional decisions at the staff level, although nursing education claims to prepare students for this level of decision-making. Rather, decisions have been made by those labeled as supervisors and administrators in the nursing bureaucracy. The question is: can these organizational structures be developed for nurses, i.e., can there be professional nursing practice within bureaucratic organizations?

Changes in Nursing Systems: Autonomy/Accountability

Changes are occurring in organized nursing services and in nursing roles in a variety of systems. The following section is a discussion of these efforts toward change as found in the nursing and health care literature. A caveat in talking about change in one system is that change in a single system affects all interacting systems; thus, changes in nursing practice affect physicians and patients, as well. This creates additional system(s) stresses and strains.

Expanding roles in nursing, as demonstrated in the nurse practitioner movement in primary care, provide an example of increasing autonomy and accountability. Nurse practitioners do physical assessments, take client histories, and frequently manage clients with common illnesses and stable chronic conditions. The expanding role implies new duties and obligations, as well as rights for the nurse in a more collaborative role with the physician. Such a change requires rearrangements in the professions of both nursing and medicine in relation to autonomy and accountability— professionally, psychologically, legally, and ethically.

A change in the administration of nursing services in the past decade has been the decentralization of authority with its consequent reduction in administrative/supervisory levels and more accountability for nursing care at the level of the patient care unit. An example of this change is the concept of the primary nurse implemented in a variety of hospital settings.[3] Primary nursing incorporates strong components of responsibility and accountability into the role of the hospital nurse. One nurse is assigned accountability for 24-hour, seven days a week care for a group of patients. Decisions for such assignments are based on patient need and the nurse's competence. The philosophy of primary nursing incorporates the concept of patient care that values interpersonal relationships, patient input, and acceptance of the advocacy role by the nurse. This includes advocacy for quality nursing care even when the primary nurse is not on duty and direct communication and collaboration with members of other disciplines caring for his/her patients. These professionals know the nurse's name and work with the primary nurse as nursing's representative for the assigned patients.

In primary nursing, contracts are made by the primary nurse with both peers and clients.[4] Duties and obligations are spelled out for all parties, a relationship of more explicit reciprocity. For example, in the primary nurse/patient contract, the primary nurse agrees to give total care each day on duty unless unusual circumstances exist, and to involve the patient and family in care. The patient agrees to share information that will affect the outcome of treatment and to participate in making his or her care plan, and also to accept responsibility for learning how to care for him or herself after discharge. In the primary nurse/peer contract, there are agreements to assist each other as necessary, to follow care plans made by the primary nurse, and to share relevant information affecting patient care.

This concept of contracts speaks to the accountability of nurses and clients and reflects the model covenant/contractual ethical relationship described by Veatch, a philosopher, for the physician and patient.[5] This model also has some relevance for the nurse/patient relationship in primary nursing. The contractual model involves two individuals or groups interacting in a way where there are obligations and expected benefits for both. Such basic norms as freedom, truth-telling, individual dignity, and promise-keeping are basic to the contractual relationship. In Veatch's model, there is a sharing of ethical authority and responsibility, a sharing of major decision-making that assures the moral integrity of both nurse and client. This is not a feature in the priestly and engineering models that Veatch describes. Briefly, the priestly model involves the imposition of the professional's values and expertise on the client; the engineering model involves the imposition of the client's values on the professional, which makes the professional simply a technician in the client's service. Supposedly this eliminates the value dimension from the decision-making process for the professional, which has obvious implications for the autonomy and accountability of the nurse, physician, and patient in the total health care system.

Significantly, several nurses with whom the author has discussed these models have suggested solutions to ethical dilemmas in nursing that are closer to the engineering model in which the patient's goals and values predominate. For example, if the patient has signed a consent form, then it is "right" to carry out whatever the form authorizes.

I have described briefly some aspects of primary nursing as an example of role change and organizational change that speaks to both institutionalized and individual nursing autonomy with corresponding accountability with client input. In this relationship, nursing autonomy is not an end in and of itself, but a means to more effective nursing care of patients/clients.

Another example of organizational change that is compatible with improving patient care and assuring professional nurse autonomy and accountability has been tested in the Iowa Veterans' Home in Marshalltown, Iowa.[6] This home provides long-term care for chronically ill and aged veterans, their spouses, and widows. Nursing care is the recognized primary service of the institution. Before 1965, care was primarily custodial. The philosophy is now one of rehabilitation, which is demonstrated in the provision of a broad range of nursing care ranging from extensive, complex nursing care to infrequent support and counselling, and includes advocacy for patients by nurses.

The nursing staff, in changing from a custodial to a rehabilitation philosophy, considered that delivery of quality nursing care depended upon the willingness and right of each nurse to practice professionally. This includes autonomy to determine and implement nursing activities and accountability for the outcomes of these acts, a collective nursing decision. The nurses used Wade's model as a guide for an organizational model that includes the identified aspects of professional practice. Several goals for professional nursing practice were established. Examples include autonomy for decision-making based on patient need, developing an environment for innovation, and change and evaluation of professional behavior only by peers. Organizational conditions necessary to achieve these goals included decentralization of authority and responsibility in the nursing department, direct access to clients, identification of all professional nurses as peers, nursing assignments in the primary nursing model, and a consensus on the philosophy and objectives of patient care.[7] Authority for decisions affecting the practice of nursing were transferred from the organizational hierarchy to the nurse peer group, a nurse collective, working with the nonprofessional hierarchy when appropriate.

Nurse autonomy and accountability are considered to be functionally interdependent under this model. One cannot exist without the other. Interdependence is demonstrated in peer consultation about patient care, peer participation in research activities to identity autonomous nurse practice, and in nurse peer review and evaluation of practice. A detailed description and discussion of the study conducted at the Iowa Veterans' Home can be found in the book, *Guidelines for Nurse Autonomy/Patient Welfare* by Maas and Jacox. Again, as in the nurse practitioner movement, one finds the dual goals of assuring improved patient care and assurance of the integrity of the practicing professional through increased professional autonomy and accountability.

Meridean Maas, former director of the funded project for testing the professional model at the Iowa Veterans' Home, provided an update for the author activities related to organizational changes and implementation of the model. A new medical director had been hired who could not understand the nurses' need for the collegial organizational structure and control over the application of nursing knowledge. He fired the director of nursing with the subsequent negative impact on the nursing staff. The director of nursing grieved this action and has been reinstated. The nursing group, which has expanded in size, is now working toward the recovery and rebuilding of what was lost by this action against the director. Efforts continue to professionalize nursing practice in this setting through development of more autonomous and accountable practice.

The above situation points up what I and many others consider to be a most profound issue in nursing and health care for providers and consumers not only in the United States but throughout the world.[8] This issue is the relationship between physicians, primarily men, and nurses, primarily women. This relationship historically and traditionally has been predominantly authoritarian—culturally, professionally, and institutionally, i.e., nursing obedient to medicine. What indeed does it mean for the integrity of both medicine and nursing when one professional group is considered to be under the supervision of another when both are responsible for application of particular knowledges—some distinct and some overlapping in practice? There have been nurses, starting with Florence Nightingale, who have challenged the control of nursing by medicine throughout the last century. But the majority of nurses have remained passive acceptors of the status quo, i.e., responders to physician orders, patient/client needs, and the demands of agency administrators and third party payers. Yet, examples can be found at the growing edges of nursing that illustrate the continuing struggle for professional practice in nursing, e.g., demonstrated effectiveness of nurse practitioners, efforts to achieve primary nursing in hospitals, and establishment of collegial decisional authority patterns in organization of nursing service as in Iowa. These efforts suggest that in terms of improved nursing care and client welfare, the on-going struggle for the control of nursing practice with concomitant increased nursing autonomy and accountability and consumer involvement in these processes does make a difference for both nurses and clients.

Looking to the Future

Changes initiated by some segments of the professional nursing community can be viewed in the context of an article by Harris Cohen, a Senior Policy Analyst with the Office of the Assistant Secretary for Health of HEW, now the Department of Health and Human Services.[9] One aspect of Cohen's discussion of the health professional's new environment could be seen as both a warning to nursing in terms of goals and priorities, and as support for present endeavors to enhance client welfare through contemporary role development and changes in organizational authority patterns.

Cohen comments on the demise of the strictly "autonomous" health professional and the movement away from professional autonomy. Independent functions are a myth owing to the growing interdependence and specializaton of function among the professions involved in health care. These realities profoundly affect the autonomy and overall responsibility of any single profession, e.g., medicine and nursing. Professional autonomy is further eroded by the incremental expansion of duties and responsibilities in certain health disciplines. Nursing is one example. Sharp boundaries no longer exist between practice professions—both a frustation and a challenge to health professionals and recipients of their care. The ramifications of this reality are personal, professional, legal, and ethical. Health care is increasingly interdisciplinary in form and substance, e.g., rehabilitation teams, some HMOs, holistic health care approaches, and community health planning efforts.

Discussion

Establishing limits to autonomy in nursing at the present time is difficult when the traditional and legally established boundaries in Nurse Practice Acts are being challenged in many states. Joint efforts of nursing and medicine through the establishment of Joint Practice Commissions at the national and state levels, consumer input as indicated in the American Nurses' Association *The Study of Credentialling in Nursing* (1979), and the systematic study of new patterns of nursing practice provide mechanisms for looking at different boundaries than have existed in the past and still persist. The question still to be addressed is, Why should there not be more autonomy in nursing practice when the responsible exercise of more autonomy by nurses has been demonstrated to improve client care?

Autonomy in nursing or any other profession is not an end in and of itself, but a means to improving client service, assuring accountability to self and others, and the personal and professional integrity of the practitioner, in this case, the nurse. More autonomous nursing practice must be balanced with efforts to achieve what both providers and consumers determine is the individual and the common good in health care. Nursing has and does provide leadership to some extent in this effort, but it is certainly not the sole determiner of even its own fate in the system where cost containment, government rules and regulations, and physicians' orders seem to be the sole arbiters of practice. Also, the past and present struggle for control of nursing practice by the nursing profession documented in the literature and in statements by those recognized as nursing leaders must be reconciled with present realities and trends discussed by Cohen, assessed needs of society, and historical and contemporary nursing documents that call for cooperative efforts with other providers and consumers. An example of one of these historical documents, which few are aware of, is the 1959 National League for Nursing Patient's Bill of Rights entitled "What People Can Expect of Nursing Service" in which nursing is recognized as an integral part of all health care and that "good nursing" requires cooperation between the consumer and providers of care, a more collegial, cooperative approach.[10] Nursing must reflect on and deal constructively with its own inner conflicts and contradictions, e.g., the goals of autonomy and accountability vs obedience to authority whether medical or administrative. Which will take priority?

After looking at aspects of autonomous practice in these past few pages it seems more productive and accountable to view both autonomy and intra- and interdisciplinary cooperation as means to the end of responsible health care for all. Three models in nursing that support just this notion have been presented to you: primary nursing, the nurse practitioners' role in primary care, and a collegial authority structure in a bureaucratic health care setting. All three incorporate more autonomy and accountability in today's complex and ambiguous world of health care, and also include the client as an autonomous, responsible person.

How we deal with the challenges inherent in the implementation of these models reflects on the maturity and accountability of both nurses and the nursing and medical professions as essential social services. If a discussion of this nature is carried out in the next century, what questions will be raised for and about nursing autonomy and accountability? Will the current questions no longer exist? They have been raised throughout the past century with different degrees of emphasis dependent on various social conditions. Nursing today has models of reciprocal relationships in which autonomy and accountability for both providers and clients are articulated aspects of practice and patterns of organization. Could they be used as models for provider/client relationships in the health care system where the boundaries of autonomy would arise from the nature of the relationships itself and the environment surrounding the relationship?

References

1. Lambertsen, E. "The Changing Role of Nursing and its Regulation." *Nursing Clinics of North America* 9: 400–401, September 1974.
2. Maas, M. "Nurse Autonomy and Accountability in Organized Nursing Services." *Nursing Forum*, XII (No. 3): 243, 1973.
3. Ciske, K. "Accountability—The Essence of Primary Nursing." *American Journal of Nursing* 79: 891–894, May 1979.
4. Ibid., p. 893.
5. Veatch, R. "Models for Ethical Medicine in a Revolutionary Age." *Hastings Center Report*, 2: 5–7, June 1972.
6. Maas, op. cit., pp. 237–259.
7. Maas, op. cit., p. 252.

8. Cleland, V. "Sex Discrimination: Nursing's Most Pervasive Problem." *American Journal of Nursing* **71**: 1542–1547, August 1971.
9. Cohen, H. "Public Versus Private Interest in Assuring Professional Competence." *Family and Community Health* **2**: 79–85, November 1979.
10. Carnegie, M. E. "The Patient's Bill of Rights and the Nurse." *Nursing Clinics of North America* **9**: 560-561, September 1974.

Chapter 5

Limits to Responsibility and Decision-Making

Some Comments

Jay Wolfson

Both Professors Engelhardt and Aroskar have addressed the issues of decision-making and responsibility in the delivery and receipt of health care services. Two basic questions emerge. First, who decides? And, second, wherein does responsibility lie? When it comes to decision-making, the answer is clear—I decide. I decide whether to use services that you give me. I decide whether to listen to the advice that you offer me. In fact, I even decide whether or not to seek services. It is all too easy for health professionals to imagine that they, in fact, make decisions that are followed by passive patients seeking their care. On the contrary, users of human service systems are not passive patients, nor are they even clients. They are conscious, decision-making consumers of services and the issue of responsibility is closely related to decision-making. How can I, a consumer of services, expect any service provider to take responsibility for decisions that I do not or refuse to make?

At the beginning of these sessions this room was divided into two sections consisting of smokers on my right and nonsmokers on my left. Imagine for a moment a situation 15 or 20 years from now when some of the people on the right side of the room contract and are suffering from various forms of cancer, probably related to their smoking. If some of these people ask to have their cancer treated and cared for, where will responsibility reside for decisions to exhaust precious resources on them when, knowing full well the risks associated with cigaret smoking today, they consciously chose to continue smoking? Can society take responsibility for their selfish decisions to continue smoking?

Now Professor Engelhardt might suggest that perhaps these people are performing a social good. By exposing themselves to an increased risk of cancer, they will, in fact, be dying off before reaching the age of eligibility for Social Security benefits, thereby saving society money; that, perhaps we should even encourage these people to smoke by giving them a smoking subsidy. But the acute care costs associated with maintaining and treating people with these cancers may far exceed the lifetime benefits received from Social Security. At issue, however, are the implications of society assuming liability for abdication of individual responsibility, and these are dangerous. When "society" becomes responsible, "crimes" literally disappear and there is only societal liability for damages, while punitive measures against transgressors become meaningless, indeed "unfair".

Historically, many of society's goals have conflicted with the personal goals of some of its members. Hobbes argued that once in a society, individuals had a duty to uphold norms and mores in exchange for protection afforded by the Leviathan (social security, Medicare, Medicaid, national health insurance) and that, in fact, the individual was bound to look out for the best interests of society by avoiding behaviors or actions that threatened its well being. When, in pursuit of vainglory, individuals transgress societal goals and deceive the benefactor Leviathan (Dante's most heinous sin) through decisions to take medical or political risk (e.g., "I will take the risk of contracting cancer by smoking") then they become threats to the well being of the system. They must stand alone and face the natural consequences of their selfishness, unaided by the strong arm and shield of the Leviathan (state-supported health care). When a system offers freedom of choice and a degree of social insurance, it is irresponsible to assume that the goals of the system become whatever individuals choose them to be at the moment.

Professor Englehardt has suggested that there are no ultimate goals in medical care. He believes that medical goals must be relativistic, contingent upon situations and needs. This assumes that both the continuum of relativity and the availability of medical care resources are endless. If the continuum of relativity were endless, then individual responsibility could easily be abdicated in the face of adverse circumstances. Were medical care resources endless, questions of responsibility would not arise.

But we do live in a very protecting system, and many members of our society have abdicated substantial amounts of responsibility for decision-making. As a culture, we have woven a cocoon of technology and bureaucracy around us designed ostensibly to ensure maximum happiness, security, and peace of mind. Is the cocoon working, or is it an illusion of security fading in the face of double digit inflation, government ineptitude, and ill-managed, questionably effective social programs?

My generation will be, perhaps, the largest cohort in history. We have been described as narcissistic, egocentric, and self-centered, with a propensity toward instant gratification. These traits may serve us well in the years ahead. As the cocoon cracks, both decision-making and responsibility become assumable when they are afforded the incentive and motivation to respond.

My young, healthy, aggressive egocentric cohorts freed from the horrible infectious and childhood disorders that struck down generations before them will have to do more than simply sympathize with the elderly and poor people in need of services today. They will have to plan for and respond to the facts of their own health care needs. In the next twenty years, it is we who will have a significant role in deciding whether we will smoke or eat well or exercise, in short, take resoponsibility for our own health through conscious, informed decisions.

The need to assume responsibility arises when a vested interest is at stake (e.g., the angry US citizen's response when the American hostages were being held in Iran.) I suggest that these conscious, vested interests must also be present with respect to the goals of medical care. In the years ahead, these medical care goals will certainly be influenced by a resource-scarce, volatile world.

But as conscious, responsible decision-makers, better educated than any generation before them, it is possible that policymakers in the years ahead will assign responsibility to transgressors at the individual as well as the corporate level. Should not the pharmaceutical firm be held responsible for exposing a large number of people to DES, making thousands of members of one generation walking time bombs of cancer? Should the large industry not assume liability for the Kepone it has dumped in the Chesapeake Bay? Should not lethargic personal health habits be discouraged through high cost-sharing provisions?

Only when individuals recognize and accept responsibility for their lives can they have the incentive to make meaningful decisions. When a personal level of responsibility is assumed, when a vested interest is accepted, and when the consequences of individual decisions are considered, the goals of human service systems can become future- and prevention-oriented. When that responsibility is abdicated, the goals can be imposed by interests not necessarily vested in the needs of the members of the system.

Chapter 6

Can Physicians Mind Their Own Business and Still Practice Medicine?

Samuel Gorovitz

My remarks here will be directed primarily to the physician as a paradigm of the health care provider. That's something of a distortion; physicians are all too often taken as paradigmatic care providers, yet there is much to be learned by looking at other kinds of health care providers. Indeed, there is much to be learned about physicians by looking at other kinds of health care providers. Further, medical care, which is what I am going to discuss, does not have much to do with health. Physicians and other health care providers should remind themselves of that each morning as they set out to work. Mainly, medical care has to do with response to illness and injury, and not primarily health. If you are serious about promoting health as effectively as you can, do not go into medicine.

The question that we will consider is: Can physicians mind their own business and still practice medicine? Philosophers have an irritating habit of not wanting to answer questions as much as to understand them better. (You know what that got Socrates in the end!) As it happens, I spend part of my time at the National Center for Health Services Research these days; my job there is that of "irritant in residence" (though that is not the way the title reads, that is the way it functions). I hope in this essay to justify that description by taking a look at the two components of our question: what is the physician's business, and what is it that is going on when physicians practice medicine.

I want to begin by examining the practice of medicine afresh; it is too familiar to us for us to understand it well without some special effort. Consider the behavior associated with medical practice. We sometimes hear physicians referred to as playing God. We hear such descriptions of their behavior when they make vital decisions. We often hear the expression "doctor's orders." But what does one get from lawyers? One gets advice, not orders. One can get into serious trouble by not following one's lawyer's advice; one can lose property, one can go to jail. It is a serious matter, yet we do not speak of lawyer's orders as we do of doctors orders. That is a clue that something is going on that needs to be looked into. And the use of the expression "playing God" is a clue.

What are some other clues? Think about names and titles. The founding documents of American democracy make it clear that this is supposed to be a society without a titled aristocracy. If you go to any quality institution of higher education and say "Mr. Smith, I'd like to speak to you after class," when it is really Dr. Smith or Professor Smith, Smith will not even notice. If you say to a clergyman, "Mr. Jones, I'd like to speak to you after the sermon," rather than "Reverend Jones," he will notice, but forgive. But you do not call a physician "Mr." unless you want to see what righteous indignation can be. And if you are not convinced that physicians have a remarkable attachment to their titles, there is a little test you can do. You can ask a group containing several physicians to perform this simple exercise. Say "Everyone take a slip of paper and write on it your name; nothing more, just your name." Most people can do that. But physicians seem to have a professional disability in respect to this task. You often find them writing down something like "Julia Jones, MD" as if a little bit of their educational history has suddenly become part of the name. What is there about the experience of having gone through medical school that makes people think it marks such a fundamental transformation that it can only appropriately be marked by revision of the name? That is a clue—this special relationship to titles, and indeed to the name itself.

So we ask, "What is going on in the practice of medicine?" Suppose we had a visitor from another planet quite different from ours, a planet in which there was no illness, no injury; everybody simply lived for eighty years and then on the eightieth birthday, peacefully and predictably, fell over dead. It would be a very different world. There would be lawyers to do wills, and morticians, but there would be no medical care. Assume someone from that planet comes to visit, and says, "What do you do?" You reply "I am a health care provider, a physician." How can you explain what that is? He says, "Why don't I just follow you around and watch the behavior? I'll see what you do." What does he see? What does the behavior look like?

You walk into a room wherein awaits a virtual stranger, and you say, "Good morning, Edward, take off your clothes. Eat these toxic chemicals. Hold still while I stick this pipe in you." Sometimes the behavior involves cutting people, sometimes even taking parts of them that you do not like and discarding them. Now, on the face of it, that would seem not best described as medical practice, but as felonious assault. The kinds of things that physicians do day in and day out are just the kinds of things that, but for very special justification, would in fact *be* felonious assault. That is something every physician should reflect on every day. Why do we permit this? How can it be that we even expect it?

There is one other aspect of medical practice that I mention because often people are quite struck by being reminded of it. It has to do with the power-related significance of waiting. When you go to a physician, you usually wait. The physician commonly has a good excuse for being late, but does not bother to give it. That institutionalized rudeness is so much a part of the conventions of medical practice that physicians are startled to have it pointed out that they are systematically rude to their patients. They have not intended to be; it comes as a revelation to them. But that is what they have been doing for years: that is how they have been taught to practice medicine.

Part of what underlies these phenomena is the fact that we are all members of that great organization, the FPA—the Future Patients Association. Everyone of us at some level, some dimly and some quite explicitly, is aware of the fraility of health and indeed the fragility of life itself. Health and life can end for us or for our loved ones at a moment's notice or with no notice at all. That is a humbling and frightening realization. And it is precisely with respect to health and life that physicians presumably have special expertise and special powers, powers of intervention that we hope will be used benevolently in our interests, but that we fear can be used or withheld to our detriment. That fear is not ill-founded. Though it is now, recently, true that on balance physicians do good, they also do a tremendous amount of harm. That is mainly because medicine is very difficult to do well. It is not primarily because of shabby medical practice, though that exists, but because even ideal medical practice is very much a guessing game. One public health service official has estimated that one hospital day in seven results from some sort of iatrogenic illness or injury. Becoming a patient is high-risk behavior. Sometimes, however, not becoming one is even higher risk behavior. So we enter into these relationships with physicians because we want something from them, a betterment of our circumstance that can be purchased only at certain substantial costs.

If you recall now that the physician's expertise is with our health and our life itself, and that the physician has special training, special powers, with respect to these goods, it is no wonder that we think of doctor's orders. For even the physician who couches an opinion in gentle suggestion speaks against the backdrop of our realization of his or her power and our fragility. Disobey your doctor's orders at your peril. And what is the risk that you then run? Not inconvenience, but capital punishment! That is the implicit threat involved in every medical recommendation, so the pronouncements of the physician induce a disposition to acquiesce on the part of the patient, both the sick patient whose regressive tendencies are well-documented, and the well patient. It is no wonder we tend to be meek in the face of medical rudeness.

If the physician's orders are compelling, it is important that the physician's autonomy, that is, the scope over which the physician has jurisdiction, be a scope with respect to which the physician really does have legitimate authority and the appropriate expertise. Imagine this transaction: your physician says to you, "I want you to take methyl prednisolone acetate, to stop eating crustaceans, to give up smoking, to get more exercise, to stop dating Mary-Sue, to vote Republican, and to sell your Volvo and buy an Oldsmobile." Somewhere along that spectrum, you have the feeling that the transaction has gone awry. With respect to the methyl prednisolone acetate, you are probably willing to go along. With respect to the crustaceans, you have a tendency to think there must be some physiological situation that makes that an appropriate prohibition. (Of course if it comes to light that the reason is that your physician is a member of a small subset of physicians who worship the crab, then you resist that suggestion as somehow transcending the scope of legitimate medical authority.) And when you are told to stop running around with Mary-Sue, voting Democratic, and driving a Volvo, you are entitled to say,

"You've crossed the border from what *is* your business to what is *not* your business."

What is the physician's business? Your health; not your politics, not your romantic affiliations, not your investments, just your health. But what is that? If the physician can give orders with respect to health, then it is important to know what health is. Englehardt has directed considerable attention earlier in this volume to the question of what the appropriate agenda may be for a physician. The World Health Organization has something to say about this matter; the preamble to its Constitution provides a definition: Health is not merely the absence of illness or injury, but a complete state of physical, psychological, and social well-being. If that is what health is, think of the consequences when the physician is considered expert with respect to your health.

A visit might go like this:

Patient: Doctor, I'm glad to see you. I'm not very well.
Doctor: What seems to be the trouble?
Patient: My social well-being is really poor. I haven't been out on a date in three months. I'm lonely.
Doctor: I think we can do something about that. I'm writing you a prescription for a particularly good computer dating service. And I am also prescribing one disco visit every two weeks. Call me in 60 days.

Is this the sort of medical care that Blue Cross will pay for? Is it even reasonable to ask that medical insurance consider such expenditures?

It is time for a literary interlude. The passage is from the novel *Oblomov*, a splendid work by the nineteenth century Russian writer Goncharov, who, as far as I can tell, lamentably is best noted for not being Dostoevsky. In this scene, we see Oblomov facing the question of whether to get up out of bed, a major decision that he ponders for nearly a hundred pages. He is visited by his physician, whose remarks illuminate our concern with the scope of the physician's legitimate authority. The physician speaks first:

> "If you go on living in this climate for another two or three years, lying around eating rich, heavy foods, you are going to die of a stroke."
> Oblomov was startled. "What am I to do? Tell me for God's sake."
> "Do what other people do, go abroad."

Oblomov here pleads that his financial circumstances will not allow that at present, and the physician replies:

> "Nevermind, nevermind, that's no affair of mine. My duty is to tell you you must change your way of life. The place, the air, occupation, everything. Everything...Go to Kissengen, spend June and July there, drink the waters, and then to Switzerland or the Tyrol. Take the grape cure."
>
> "The Tyrol. How awful...But what about the plan for the reorganization of my estate? Good heavens, Doctor, I'm not a block of wood you know."
>
> "Well it's up to you; my business is simply to warn you. You must be on your guard against passions, too. They hinder the cure. Try to amuse yourself by riding, dancing, moderate exercise in the fresh air, pleasant conversation, especially with a lady, so that your heart is only moderately stimulated and by pleasant sensations."
>
> "Anything more?" asked Oblomov.
>
> "As for reading or writing, God forbid. Rent a villa, facing South, plenty of flowers, lots of music, ladies around."
>
> "What sort of food?" asked Oblomov.
>
> "Avoid meats, fish, fowl, also starchy foods and fat. You may have light boullion, vegetables, but be on your guard, there is cholera almost everywhere now. You must be very cautious. You may walk eight hours a day."
>
> "Good Lord," Oblomov groaned.
>
> "Finally, spend the winter in Paris. Amuse yourself in the whirl of life. Don't think. Go to the theatre, to a ball, a masquerade, out of town. Surround yourself with friends, with life, laughter."
>
> "Anything else?" Oblomov inquired, with barely concealed annoyance.
>
> "You may benefit from sea air. Take a steamer to England, from there America. If you carry this out exactly..."
>
> "Certainly, without fail," said Oblomov sarcastically.
>
> The doctor went out leaving Oblomov in a pitiful state. He closed his eyes, put both hands to his head, curled up in a ball in his chair, and sat there neither seeing nor feeling anything.

What has the doctor told Oblomov to do? To become someone else! He has taken the World Health Organization's definition of health seriously. There is no dimension of life that cannot affect one's health. There is no dimension of life that cannot affect one's physical, psychological, or social well-being. If that is what health is, and if physicians can give orders with respect to health, then physicians can give orders with respect to every dimension of life.

Have the medical schools prepared them adequately for such responsibility? Of course, the answer is "No," and must be "No." There is a limit to the competence of the physician, and the scope of legitimate medical autonomy can extend no further than the scope of medical competence. Moreover, the limits of competence merely set an outer constraint on medical autonomy. Indeed, the total constraint is more severe than that, because being right about what is best for you does not by itself entitle me to order you about. Since smoking seems to be the theme here, I will illustrate with that example. I have no doubt that I am right in thinking that smokers are among the most self-destructive of drug addicts, as well as being all too often among the most uncivil, but it does not follow from my being right that I have any entitlement to try to prevent smokers from smoking. I can quite well prevent them from doing it at me, or in my home or office, but I mean on their own turf. I would not presume to do that. Being right about what is good for someone else is not sufficient to justify giving them orders. So the scope of competence is broader than the scope of legitimate authority, even though the scope of competence is quite narrow.

Let us look at two illustrative cases. (One must have real cases before any medically sophisticated audience, otherwise one is not credible. These are real cases.) First: I visited a hospital in San Francisco recently: one of the issues in greatest dispute was how to make decisions about long-term neonatal intensive care of newborns in the range of 600 grams. There were a number of cases that the nursing staff believed to be clearly hopeless, while the medical staff believed that maximum effort should be made to the last second to save the child. Such efforts incur tremendous costs—psychic costs to the nursing staff, the parents, possibly the child itself, and economic costs. What is there in medical training that equips the physician to make a well-informed judgment about what sort of cases are appropriate for valiant intervention and what sorts not? As far as I can tell, nothing whatever. But the power structure in the hospital in such that the physicians typically make the decision. That is an example in which the authority exceeds the legitimate scope of expertise.

One cannot help but be reminded here of Englehardt's observation about the esthetic attraction of using sophisticated skills. Two people, both of whom know chess and checkers, are likely to play chess; they rarely play checkers. Skills that have been acquired at substantial personal cost are skills that people like to use. People who can do sophisticated things like to do them.There is an intrinsic payoff in satisfaction. State-of-the-art medicine is very sophisticated, and people who can do it often find it a very beautiful thing to be doing. We should bear that in mind when we think about the motivation behind a lot of medical care.

Here is the second case. An octogenarian approaching death with end-stage cancer was very uncomfortable despite having been given analgesic medication. Her son asked the physician if there were not some way to increase her comfort during her final days. The physician said, "I'm sorry, but we're doing everything that can be done. But she can't last much longer; it is a matter of days at this point."

The son, being a philosopher, went away thinking about what he had been told, and made it across the street before turning back to see the physician again. "I'm sorry" he said. "That just isn't good enough. I want you to explain why."

The physician was ideal. He said, "Don't apologize at all. I want you to feel free to ask me questions at any time, here at the office or at home. Ask new questions that occur to you, or ask the same question over again, because sometimes people need to do that, too. Here's my home number; don't feel shy about calling me there if you think it will help." And then the physician gave a little lesson in clinical pharmacology.

"How much medicine we can give in any particular case is determined in part by a recognition of the fact that anything that's a drug is also a poison. If it is a drug at all, it is toxic, and there is a delicate trade-off. What relieves the pain also suppresses the respiration, and invites an infection. We have to consider total body weight and basic strength. The primary risks are that we can suppress respiration to a point that will itself kill the patient, or will invite something that will kill the patient, and there are problems of increasing dependency, of needing the drug at shorter and shorter intervals and becoming increasingly addicted."

The son thanked him and went about as far as the elevator, then he turned back. He said, "Okay, I think I understand now what is going on. Now, I want you to understand that I'm the last person to advocate that what this society needs is terminally ill octogenarian pushers out on the street. I understand that as a general convention of clinical practice it makes sense to be cautious in the use of drugs. But in this particular case I don't see the sense; I don't see the problem of possible addiction as having any bearing on this case. And as far as suppressing respiration is concerned, where is it written that the malignancy has an entitlement to be the cause of death, as if somehow the physician has an obligation to keep the patient alive until the malignancy has had time to cash in its claim? Why is it a bad thing if she dies of a drug-induced respiratory infection, if she's more comfortable in the meantime?" The physician had no ready answer. The son continued. "I see two conflicting values here, the prolongation of life and the relief of suffering. I see a convention in medical practice of how to respond to such a conflict. I see a physician who knows how things are usually done, who knows why they are done that way, and who recognizes that it is best that they usually be done that way. And I see those conventions and principles being applied here to a case to which they just don't apply. Prolongation of life in general is more important than relief of suffering, but in this case those two values in conflict should not be weighed in the usual way. In this case, relief of suffering is a more important value than the prolongation of life. Not only have you assumed which value was the more important in this case, and imposed that assumption on your patient, you've done it without being aware that you were doing it at all."

The physician replied, "That's entirely right." The doctor then changed the prescription, and the patient became more comfortable. She died shortly thereafter. I do not know what she died of nor whether she was addicted to anything at the time, and the reason I don't know is that no one thought these were interesting questions.

What has happened here? It is important to note that this is no dramatic, front page case. It is not even a case that involves a vital decision in any significant sense. It is really a rather mundane piece of medical business. What happened was that ethical decisions were being made by the physician unwittingly, and imposed on the patient. Making such a decision was, in a sense, none of the physician's business. How to implement the choice, however, is indeed the physician's business. What the physician did in this case was to overstep the bounds of his competence, of his expertise, of his legitimate authority, not out of maliciousness, not out of an imperialistic attitude, not out of any desire to play God—anyone who has been at all close to medical practice must realize that in such circumstances the last thing that physicians are doing is *playing* anything—but because of a lack of moral sensitivity. In cases such as this, incidentally, the nursing staff often has more insight into what is happening in the physician's behavior than the physician has.

So, my answer to the question, "Can physicians mind their own business and still practice medicine?" is "No, they can't, because the practice of medicine is essentially the minding of somebody else's business." What is crucial for practicing medicine well is that those who are doing it recognize that, and learn to tell the difference between what is their business and what is somebody else's. The practice of medicine, in short, requires minding somebody else's business and not mistaking it for your own—recognizing the limits of medical authority.

It is partly because of this that I emphasize the point about felonious assault. Some physicians recognize that much of their behavior is assaultive, and they recognize that the justification for it is that because it is in the interest of the patient, the patient grants a waiver, so to speak, legitimating specific medical interventions. But too many physicians think that what justifies medical intervention is their status and stature as physicians. Yet the license to practice medicine does not entitle the physician to do anything to anybody without special justification for that act of intervention. Specific medical interventions, however assaultive, can be justified. But there is no general justification. Interventions must be individually justified case-by-case, patient-by-patient, intervention-by-intervention.

What the physician is called upon to do involves, in an essential way, minding somebody else's business, yet physicians are often not well aware of the distinction between what is their proper business and what is the patient's. This point is not made in medical education very effectively. So part of what I want to stress is that one reform that is needed in medical education is for physicians to be trained to have a greater awareness of the distinction between what is their proper business and what is not.

I will conclude with one last case. Recently, in a hospital in the West Coast, a man was dying. His blood pressure had practically vanished and his brain wave was essentially flat. One resident said to a nurse, "I want to do a Swan-Ganz line, please bring the materials." The nurse talked to another nurse, and they decided they would not comply because there was clearly no potential benefit to the patient from this procedure. At their peril, they refused. But the resident and another more junior resident got the materials themselves, and initiated the procedure about forty minutes before the patient died. During this time the patient's family was in the hospital, but did not have access to the patient. Had they wanted to be with him for the last moments they could not have. (I do not know whether they did want to.) The nurses were upset at what was clearly an instance of doing an invasive procedure on a patient it could not help, simply in order to practice doing the procedure. They became more upset when it came to their attention later that the patient's account was charged two hundred dollars for the Swan-Ganz line. If the procedure is done, there is a charge of two hundred dollars, and somebody must pay.

It took quite a while, and some courage, and some threats; finally the nurses brought it about that the residents were criticized and the charge was removed. Those physicians had wanted to learn how to do a Swan-Ganz line, and thought that no better way exists than to do it when there is a chance. But they had not been taught anything appropriate to the making of the decision about what kinds of practicing are acceptable and what kinds not. They had the power to make that decision, but they had no education, no wisdom to enable them to see what they did as unwarranted.

It is important to recognize that the victims of inappropriate medical practice are not always patients. They can be insurers, they can be the patient's family, they can be ancillary staff—people working for and with physicians, they can be those who are connected in any way with the episode. So the question of the legitimate scope of medical authority is an important and pervasive question, one we need to think about carefully.

Note

1. Portions of this essay are drawn from the author's forthcoming *Doctors' Dilemmas: Moral Conflict and Medical Care,* Macmillan, 1982.

Chapter 7

Allocating Autonomy

Can Patients and Practitioners Share?

Sally Gadow

On the surface, conflicts between professional authority and patient autonomy seem almost obsolete. After great expenditure of federal funds and local time, health care is becoming so humanized and patients so educated that—instead of meeting each other halfway—they have all but exchanged places. As professionals become increasingly human and patients increasingly professional, little turf remains that belongs exclusively to one or the other. Health care would seem to be well on its way to a situation of "shared autonomy."

A closer look, however, suggests otherwise. Neither consumers nor providers are well on the way to sharing autonomy, in spite of appearances. At the policy level, physicians retain exclusive control, for example, of patients' use of prescription drugs, while patients, from their side of the fence, litigate increasing control over physicians. At the clinical level, practitioners anguish over patient's noncompliance, while patients study texts such as *How to Manage Your Doctor*. The very concept of noncompliance betrays the professional's concern about patient autonomy, and the growing movement in patient assertiveness suggests the patient's view of professional autonomy.

It is my premise that the issue of who should control whom in health care is a function of a specific philosophy, namely, that in which the autonomy of one party is assumed; that of the other suspended. However, since this familiar view of medicine, fondly termed paternalism, already has been soundly thrashed by its critics, I shall direct my discussion to an alternative philosophy.[1]

A radical reconceptualization of health care is called for if we are to extricate ourselves from the assert-and-counterassert struggle. Wresting power from the provider and transferring it to the consumer may achieve cooperation, but, as every child on the playground learns, cooperation that is achieved by force will always depend upon who happens at the moment to be stronger, not who is "in the right." For pragmatic reasons, then, if not for ethical ones, professional autonomy ought not be "taken by force," stormed like a citadel in order that the consumer's flag finally be raised over the smoking ruins. What is called for instead (unfortunately, for those interested in a quick coup) is a complete philosophical reorientation of the health professions.

Self-Care Philosophy

Nothing less than a new philosophy of health care is required. The view that I shall develop and defend here is a view that can be called the self-care philosophy of health care.

First, a word of caution against a possible confusion. "Self-care" is a term usually met in the context of the lay revolt against professional care. Viewed in that way, it is no more than a guerilla tactic in the attempted overthrow of medicine. It is billed as an *alternative to* professional care *rather than a philosophy of* professional health care.[2]

What is meant by self-care as a *philosophy* of health care? Specifically, how does such a philosophy advance us beyond the struggle for control between patients and practitioners?

The view of self-care I am proposing is this: that professional health care have as its purpose a positive assistance to individuals in the development and exercise of their autonomy in health matters. Self-care, as I am using the term, refers to actions personally selected and initiated, and thus freely performed by an individual in the interest of promoting that person's health as he or she construes it.

Two assumptions are crucial here. The first, which I shall not elaborate, is that self-care is an essential expression of individual freedom. Consequently, it is an aspect of human existence that, ethically speaking, ought to be enhanced. The second assumption is that self-care must express the individual's uncoerced self-determination if it be care that is, in truth, personally initiated and freely performed.

Clearly, "self-care" in this sense means much more than is usually meant by the term. On this view, self-care does not make the patient a "physician extender." That is, self-care is not carrying out the doctor's orders at home. But neither does it make the clinician irrelevant. That is, self-care is not a form of "alternative health care" that uses the professional—if at all—strictly as a consumer's guide.

What then *is* the place of the professional on this view? The concept of self-care that I am proposing involves two dimensions, and in both of these the professional provides a necessary assistance. These dimensions we can distinguish as *action* and *agency*. Self-care *action* is the health behavior that is performed. Self-care *agency* is the process of self-determination, the decision making with respect to the choice of behaviors and, more fundamental still, the determination of an individual's own values and concept of health upon which to base health care decisions. For all health behavior that has not become automatic, conscious choice necessarily is involved, a decision must be made. That decision is not "Can I do it?" but "Shall I choose to do it? Does the behavior represent a value for me? Does it facilitate my health as I understand it?" The conscious addressing and answering of that basic question about personal health and values is the exercise of self care *agency*, the process of self-determination.

Health practitioners have long been familiar with patients' needs involving self-care action, for example, in chronic illness and rehabilitation. Needs involving agency are not as familiar, especially since many professionals assume that agency is so impaired in illness that heroic measures are always indicated, namely, the substitution of professional authority for patient decisions. Within the self-care framework, however, self-determination is central. Coerced, enforced, or directed health behaviors may be conducive to health (professionally defined), but they *are not self-care actions* because they *do not originate in self-care agency*, that is, in the patient's own free decision.

Two forms of agency can be distinguished in self-care: one, the exercise of moral autonomy; the other, the formulation of clinical judgment. I shall argue that they are finally inseparable in self-care.

Self-Care Agency: Moral Autonomy

The philosophy I am describing involves the obligation that professionals assist patients with agency as well as action. This means that, while practitioners are obligated to act in their patients' best interest (as they are in any philosophy of health care), here it is the patient and not the professional who decides what "best interest" shall mean. This is not only morally entailed by the self-care philosophy. It is also required by the fact that "best interest" for a patient —that is, patient self-interest—cannot be professionally defined. A definition of self-interest, or patient's best interest, that is determined by anyone other than the patient is not only unethical on this view, but in fact impossible.

Exactly what is the difficulty in ascertaining for a patient the course of action that is in that person's best interest? Is not that very determination the essence of health care?

Some of the difficulties in determining "best interest" for patients are by now familiar. The difficulty of defining "health" has been widely discussed.[3] As a result, although we have not yet reached the point of complete relativism, we at least acknowledge multiple levels on which to meaningfully define health and illness, from the molecular to the political. But that diversity among concepts of health is only part of the difficulty in determining the self-interest of patients. The other difficulty is that the concept of "self" is as elusive as that of health. To illustrate, there are at least three dimensions that must be examined in determining one's own best interest and thereby (subjectively) defining the self. These dimensions are the inner complexity, the outer boundaries, and the uniqueness of the self.

Complexity

For what *part* of the self should one care? The psychological? Intellectual? Physical? Spiritual? Seen from the outside, the self may seem to be a simple unit. In reality, it can be dismayingly complex, in spite of personal and professional efforts to subdue dissident elements. Consequently, certain aspects of the self may receive care at the expense of other parts. A decision that seems to ignore health needs may be a reasoned choice to care for the part of the self that the individual values most. A woman may sustain an "unhealthy" drug dependence in order to have available all of her energy for dealing with an overwhelming grief. She chooses to care for the part of the self that is suffering and in effect may be using the drug as a responsible form of self-care. The objection will arise that this is not true self-care, but is, at best, misguided; at worst, irresponsible or abusive. But a health care philosophy cannot have it both ways. Either a decision is valid because it is based upon self-determination and represents the individual's own definition of self and self-care, even if that definition conflicts with objectively established standards. Or, objective norms are imposed as the appropriate standard of care, in which case, health care is based not upon a philosophy of self-care, but upon the tradition of normative care.

Boundaries

A second difficulty in defining self-interest, in addition to the inner complexity of the self, is the extensiveness of the self. Is a human being essentially a free-standing, atomic entity, a self-contained unit? Or, does the individual exist primarily in relation to others? If the latter is the case—as it is whenever we choose to define ourselves through our relationships—then self-care would be an attending to the union rather than the individual, the "we" rather than the "I." Thus, a father might choose to give up a valued position in order to spend the first years of his child's life developing a close relationship with the child. He chooses to live essentially in a union with another person rather than as an independent entity, at least for a time. He has not chosen to sacrifice his own interests to those of the child. Instead, he regards the child and himself not as two self-contained individuals (who would in that case indeed have competing interests), but as a union.

Uniqueness

A third difficulty with any objective definition of self and self-care is that not only is the self complex and at times wider than a solitary individual, but it is unique. The fundamental health needs that self-care addresses may be universal., but they are perceived and understood in (often vastly) different ways by individuals. Therefore, no universal standard of self-care is applicable. Any generalization compromises uniqueness, since it assumes that individuals are more alike than different, that is, that uniqueness is not a central, but a peripheral feature of human experience. That assumption has merit, of course, if we are concerned with establishing general norms of health conduct and achieving compliance by imposing those norms upon persons. But the adoption of health behaviors by an individual in compliance with standards governing *persons in general* does not constitute *self*-care. The basis of self-care is self-determination, freely exercised choice expressing the individual understanding and values of the patient rather than the normative expectations of health authorities. Paternalism, the coercing or forcing of persons to act in their (externally and generally defined) best interest, is therefore—to repeat an earlier point—not only morally, but logically impossible on this view. Because each self is unique, a decision about the best interest of an individual cannot be made by anyone except that individual.

The uniqueness of the self is especially significant in relation to persons who, we readily admit, differ from the "normal", but who—for that very reason—are allowed *less* self-determination than other persons. I alluded to this issue when I observed above that many practitioners feel that the need for assistance in one's health care is tantamount to the need for professional decision-making. According to that belief, the greater the health need, as in severe injury or life-threatening illness, the greater the justification for substituting professional judgment for patient self-determination. In extreme situations, of course, patient decision-making is often impossible because of uncontrolled pain, unconsciousness, disorientation, lack of information, or other reasons. But in non-acute situations the reasons are not as clear, that is, in cases of chronic physical, psychological, or intellectual handicap, early childhood, advanced age, and so on. Persons in all of these situations deviate so far from "normal" self-care capacity that we have even developed clinical specialities to address their needs. Yet, for persons in these very groups, we seem to know less rather than more about their distinctive self-care needs, in particular, their need for assistance with self-care *agency*. It does not seem an exaggeration to say that the more unmistakably unique and nongeneralizable a situation is, such as the self-care of a multiply handicapped child, the more likely we are to tailor our assistance to general standards rather than to decisions arising out of the uniqueness of the individual.

It must be acknowledged that for many reasons, including limited resources, self-determination cannot be fostered in certain complicated situations. There, the external approach must be taken if health in any sense is to be maintained. However, it should be clear that in those cases, and even in cases in which patients willingly practice the behaviors prescribed, self-care has not yet been achieved. Patients who have been taught only to observe *standard* health care practices have not yet been helped to practice self-care, that is, care based upon *individual agency.*

The question now must be addressed, how does the professional assist patients in the exercise of self-care agency? That assistance involves the professional in helping individuals become clear about what they want to do, by helping them clarify their values and beliefs in the situation and, on the basis of that clarification, to reach *decisions that express their understanding* of themselves and the situation. That assistance is needed because clarification can be the most difficult exactly when it is the most urgent, when a situation arises that calls into question beliefs and values that are often the most deeply embedded in the person's life, and thus perhaps are the least defined or developed.

The role of the professional in self-care then is not merely to *protect* patients' freedom of self-determination against institutional encroachments, inadequate information, or professional prejudices. It is far more than that. It is *active assistance* to individuals in their exercise of self-determination. It entails participating with patients in discussion that has as its mutually agreed upon purpose the discernment of the patient's view and values and the examination of health care options in the light of those.

In the self-care framework, a practitioner is not relieved of ethical deliberation simply because it has been ascertained that the patient "owns the problem." On the contrary, the professional participates in the deliberations as fully as if he or she "owned the problem." The crucial difference is that that participation serves to assist the patient rather than the practitioner. Thus, as much preparation and proficiency in ethical reflection is required of the professional on this view as on the paternalism model where the professional shoulders all of the decision-making. More, in fact, is required, since the practitioner is concerned with a problem involving not his or her own values, but the patient's, which at the beginning may be undefined with regard to health.[4]

Self-Care Agency: Clinical Judgment

The discussion of agency thus far has centered around freedom, self-determination, and values—all of them moral concepts. I have attempted to establish the *moral autonomy* of individuals in the self-care framework.

Now, a further issue arises in connection with agency. Even if we grant the *moral* agency of patients in regard to decisions about values, can we possibly suppose that patients should exercise autonomy in essentially clinical matters such as the formulation of diagnosis, the determination of etiology, or the selection of treatment? To phrase the problem differently, even if patients ideally ought to have as much autonomy as practitioners, can they possibly have as much knowledge?

The self-care philosophy of the patient as agent means that the professional incorporates the patient's personal, unique perspective on the situation as not only ethically, but clinically, primary. The patient's view must be valued as being as credible and as crucial as the clinician's approach. The reason for this is not the imperfect state of medical science. The reason for the validity of the patient's perspective is the phenomenological difference between viewing an experience as the subject or agent of that experience (as the patient views his or her illness) and viewing an experience objectively (as an object that is not largely constituted by the viewer, the clinician). Philosophical phenomenology has shown that the subjective activity of giving sense and meaning to one's complexity is at least as important as the application of objective categories in trying to understand a situation.[5] Inevitably, the patient brings his or her own perspective to bear on the problem at hand, and within a philosophy of patient agency, that perspective becomes an essential, not an optional, corollary to the professional's view in their combined clinical judgment.

It is not intended that the patient become an amateur clinician. Instead, it is proposed that the patient–practitioner relationship combine and make mutually enhancing the two different perspectives represented; either one alone must be seen as inadequate, one-sided.

To accept the validity of the patient's input into the formulation of a clinical judgment, we only need to accept the essential complexities in the experience of illness. Earlier I suggested some of these complexities in relation to the definition of "self." To compound the difficulty—or to add to the richness, depending upon one's view—there may be as great a complexity attending "the body." We can distinguish at least three modes of experiencing one's body, and a brief description of these may illustrate my point. Two of these modes phenomenologists refer to as the lived body and the object body; the third, I shall call the body as subject. They can be understood as different levels of the relation between self and body.[6]

Lived Body

As this level no distinction is experienced between self and body. The body is not that with which I act; it *is* my acting. Conversely, it is also my immediate vulnerability to the world acting upon me, my direct relation with the world, unmediated by any awareness of the body "over against me."

Object Body

When pain or incapacity disrupt the immediacy of the lived body, a distinction emerges between self and body. The body is then an encumbrance, an impediment that the self must take into account. The body is that which the self must either learn to control or be controlled by it. In contrast to the *lived* body unity, the body is now *object*.

"Object" refers here, not to the abstract theoretical body that the anatomist studies, but to the existential otherness, the concrete "thinghood," of the self. In this sense, the person who understands nothing about the body as neuroanatomical object experiences the existential object body in using the arms to lift a paralyzed leg.

Subject Body

In illness, the body is not only object, but enemy. The body mutinies, the self is overthrown. However, if we suspend the object body framework, a different phenomenon is possible: in illness the body insists, not that the aims of the self be surrendered, but that the body's own reality, complexity, and values be acknowledged. The acceptance of that reality as valid is the experiencing of the *body as subject*. That is, when the body is experienced as subject, it is considered to be part of the self with the same intrinsically valid claims as any other part of the self.

The contrast between object body and subject body can be illustrated by the difference, for example, between experiencing a phenomenon such as shortness of breath as limitation upon activity that might be alleviated with proper conditioning, or as an expression— a symbol—of the body's own perspective on the value of activity, speed, endurance. The slowing of the aging body thus can have either negative or positive meaning, as either the inability to move as quickly as the object body was once conditioned to do, or as the new capacity for "opening time," that is, for exploring experiences "not in their linear pattern of succeeding one another, but in their possibility of opening entire worlds in each situation."[7] By positive meaning, then, is meant a consideration of the particular phenomenon, such as slowing, as the body's expression or symbol of its own meaning.

The terms in which the subject body is perceived and understood cannot be the language of physioanatomy with which the clinician approaches the body. The classifications of science presuppose a body whose reality already is fully established or is developing along a typifiable course. But that is not the kind of reality that characterizes a subject, a self. Thus, categories that comprehend only a fixed, predeterminable object are inadequate for perceiving and interpreting the development of the body as subject.

In effect, scientific language is inadequate for describing the subject body because it is designed to express only a finite reality and meaning. However, *as self*—that which develops its own reality and meanings—the body is infinite. It is not of course infinite *as object*, able to transcend space and time. Its infinity is the infinity of self, that which transcends fixed determinations.

With this brief excursion into the phenomenology of the body, I am suggesting that even one of the body's realities, namely, the patient's own experience of the body, is not a simple affair. Now add to this complexity other perspectives toward the patient's body—notably, the clinician's view of the body as scientific object—and it should be apparent how elaborate the process of clinical judgment must be, to do justice to all of the realities present in a single determination about care. This does not mean that clinical judgment in the self-care framework is an unhappily mixed marriage of competing views. On the contrary, my point is that only within self-care practice *as a collaborative enterprise* between patient and professional *can either one* exercise clinical judgment, that is, informed decision-making about diagnosis and treatment. Thus, it is as important that the clinician's perspective be incorporated into the judgment as it is that the patient's view of the problem be respected and incorporated as phenomenologically valid. Finally, however, in the task of their together creating as comprehensive an understanding of "the whole person" as possible, it will be the patient who determines which findings are to count as thematic for the clinical judgment.

Conclusion

The discussion has indicated one way in which a complete reconceptualization of health care might proceed. In suggesting a fundamentally new philosophy of health care, I have not proposed self-care as an alternative to professional care, nor as a means of subverting the medical establishment. In both of those cases nothing substantive can be accomplished as long as the consumer–provider antagonism persists. Giving the power "to the people" may put the consumer in control. But my point is that any health care philosophy in which considerations of control and power are pivotal *cannot* in principle provide a framework for sharing autonomy, and—for reasons both ethical and medical—the sharing of autonomy as I have described it in terms of self-care is the preferred alternative to both of the two power extremes, consumerism as well as paternalism.

In the philosophy I have proposed, the professional actually contributes more than in a paternalistic framework. At the same time the patient exercises even greater self-determination than on the consumer model.

First, the professional contributes more than on the paternalist model because the clinician is not concerned with reaching a unilateral clinical judgment *about* a patient, but instead engages *with* the patient in an endeavor to reach a joint understanding of the situation —an understanding that expresses the patient's values and concepts of the body, the self, and health, as well as the practitioner's own perspective on these same issues. This unquestionably is an endeavor far more involving intellectually and ethically than simple decision-making *on behalf of patients*.

Second, patients exercise greater self-determination on the self-care model I have described than on the consumerism model. They are frankly supported instead of challenged by the clinician in their exercise of autonomy. In fact, patients who claim to want professionals to make all of the decisions may find themselves gently drawn by the practitioner into the clinical deliberations in order that their freedom of self-determination not be waived for trivial reasons, that is, out of habit or deference. In short, the aim of the health professional on this view is to actively assist patient self-determination. Given the difficulties inherent in clarifying one's view of the self, one's relationship to the body, and one's concept of health, that assistance from the health professional cannot fail to enhance the degree of self-determination that is exercised by patients.

Notes and References

1. The usual arguments against paternalism are grounded in the principle of individual autonomy. Ivan Illich and Thomas Szasz are probably the most vociferous defenders of that principle in the health care context. A different approach is developed by Allen Buchanan ("Medical Paternalism," *Philosophy and Public Affairs*, (No. 4), 1978), who develops an antipaternalist critique that meets the paternalist "on his own ground" without recourse to rights-based arguments.

2. For a discussion of the potential role of the layperson as self-provider of primary care, see: Levin, Lowell S., Katz, Alfred H., and Holst, Erik. *Self-Care: Lay Initiatives in Health:* New York, Prodist, 1976. The authors, though suggesting that "a new lay–professional partnership may emerge," do not commit themselves philosophically to such a partnership.

3. See, for example, Englehardt, H. Tristram, Jr. "The Concepts of Health and Disease," *Evaluation and Explanation in the Biomedical Sciences*. Englehardt, H. Tristram, Jr., and Spicker, Stuart F., eds. Dordrecht, Holland, Reidel, 1975; Callahan, Daniel. "The WHO Definition of 'Health'," *The Hastings Center Studies*, Vol. I, No. 3 (1973); and Boorse, Christopher. "On The Distinction Between Disease and Illness," *Philosophy and Public Affairs* 5 No. 1 1975.

4. In "An Ethical Model for Advocacy: Assisting Patients with Treatment Decisions," I propose guidelines for professional assistance to patients in their deliberations about values and treatment options (in *Dilemmas of Dying: Policies and Procedures for Decisions not to Treat*, Wong, Cynthia B., and Swazey, Judith P., eds. G. K. Hall, Boston, 1981).

5. See especially Merleau-Ponty's discussion of the body in *Phenomenology of Perception* (Colin Smith, trans.), New York, Humanities Press, 1962.

6. For a more detailed discussion of these levels, see my "Body and Self: A Dialectic," *The Journal of Medicine and Philosophy*, 5, No.3, September 1980. For an analysis of the lived–body in particular, see: Schag, Calvin O. "The Lived Body as a Phenomenological Datum," *The Modern Schoolman* 39 (1961–62), 203–218.

7. Berg, Geri, and Gadow, Sally. "Toward More Human Meanings of Aging: Ideals and Images from Philosophy and Art," in *Aging and the Elderly: Humanistic Perspectives in Gerontology*, Spicker, Stuart F., et al. (eds.). Humanities Press, New York, 1978, 83–92.

PART II
Refusing/Withdrawing from Treatment

Chapter 8

The Right to Refuse Treatment

A Critique

Thomas Szasz

I

The phrase "the right to treatment" became popular during the 1960s. As soon as it did, I denounced it as pernicious nonsense, contrived by public interest lawyers who fancy themselves to be civil libertarians, but who are, in fact, the bitterest enemies of civil liberties.[1]

The concept of the right to treatment was based on a malicious proposition. Ostensibly intended to prevent the abuse of "warehousing" involuntary mental patients in state hospitals (that is, confining them without treatment), the doctrine of the right to treatment compounded this evil by adding to it the evil of involuntary psychiatric treatment (mainly in the form of forced drugging). Since involuntary treatment had long been a part of the psychiatric armamentarium, a formally articulated "right to treatment" for mental patients was patently a tactic for strengthening coercive psychiatry: it granted persons who did not want to be patients a "right" to a treatment they did not want, while it continued to deprive them of the right to reject the sick role.[2]

Predictably, the doctrine of the right to treatment quickly led precisely to the sorts of abuses one would expect. Instead of correcting the errors that spawned these abuses, the mental health reformers added insult to injury by creating the doctrine of the right to refuse treatment. According to their supporters, these theories and tactics complement and correct one another.[3] I disagree. Both of these doctrines are inherently deceptive. Both are inimical to the liberty and dignity of individuals categorized as mental patients. Previously I have registered my opposition to the concept of the right to treatment.[4] Now I want to register my opposition to the concept of the right to refuse treatment.

II

Life is a very complicated affair. Children are unlike adults. The poor are unlike the rich. The unconcious patient presenting a medical emergency differs from the concious patient seeking an elective medical intervention. Obviously, no single arrangement or model for dispensing medical care can work in all circumstances. Without entering into the complexities of this matter any further than necessary, I shall analyze the concept of the right to treatment as it appears in three currently prevalent types of medical arrangements—namely, paternalism, communism, and capitalism (using these terms, so far as possible, descriptively rather than pejoratively).

Let us begin with paternalism, which is the natural model of medicine. Doctors are, on the whole, paternalists. They are trained to be paternalists. Much of the time they *have* to be paternalists to do their job.

Seriously ill people are, interpersonally and politically, much like children. They suffer pain, are fearful, disoriented, unable to comprehend what is wrong with them, and intensely dependent on those who offer or promise to help them. They may be unconscious. Clearly, the more severely disabled and helpless persons are, the more likely it is that they will perish unless they are actively cared for—transported to places of safety and care, administered treatment, and so forth. Such persons may be physically unable to give consent to treatment. In this sort of situation, what is expected of the doctor is care and compassion—and of the patient, cooperation and trust.

This model is dear to the hearts of doctors as well as many patients or would-be patients. Both groups believe, I think correctly, that there is no other way to treat small children and unconscious adults. They also believe, I think incorrectly, that many other types of persons—for example, the elderly, the poor, the mentally ill—resemble children and unconscious individuals so closely in their inability to govern their medical destiny that they, too, ought to be treated along paternalistic lines.

Historically, psychiatrists have loved this model. Accordingly, they have distinguished, first, between those who are mentally healthy and those who are mentally sick; and secondly, between those who are mildly ill (the neurotics) and those who are severely ill (the psychotics). It is of the very essence of the latter distinction that the severely mentally ill differ from the mildly mentally ill (or the normals) in their inability to assume responsibility for their own welfare, including their need for psychiatric hospitalization and treatment. Actually, a core meaning of the term "psychosis" lies in its reference to a "psychiatric illness" of such a nature and severity as to justify—indeed, require—imposing treatments on patients that, were the patients not psychotic, they themselves would desire, choose, and gladly submit to.

This is an internally contradictory proposition, but the contradiction is psychiatry's making, not mine. Because the person is psychotic, that individual cannot ascertain and act in his or her own best interest. If the person were not psychotic, he or she would choose the treatment the psychiatrist seeks to impose. But if the patient were not psychotic, no treatment would be needed. There is no way to resolve this paradox—one that psychiatry shares with other paternalistic arrangements of decision-making, such as parenthood, colonialism, and so on.

To illustrate and document the prepotent role of the paternalistic model in contemporary American medicine, I offer the following statement of the American Medical Association's view of the correct relationship between physician and patient with respect to the patient's access to medical information:

> It is our position that the right of a patient to medical information from his physician is based upon the fiduciary relationship which imposes a duty to act in the best interest of the patient. ... It is our position that the physician has both a right and duty to withhold

information in circumstances in which he reasonably determines that it would not be in the best interest of the patient.[5]

The AMA thus recommends that even competent adults should be deprived of correct medical information about their own conditions—if their physicians decide that that is in their best interest. What makes this rule quintessentially paternalistic is that the physician who possesses the information is also the expert who decides whether it is good or bad for the patient to have it.

The American Psychiatric Association's (APA's) commitment to a paternalistic model of the doctor–patient relationship is so deeply entrenched and so well-known that it hardly requires documentation. A recent example of the Association's devotion to this principle is its objection to so-called patient package inserts intended for the users of psychiatric medications. *Psychiatric News* (the Association's official newspaper) reported this development as follows:

> The APA has objected strongly to the Food and Drug Administration's proposal to require patient drug information—so-called patient package inserts or PPIs—to be dispensed with all prescription medications. ... The APA recommends that any regulations incorporate several principles:
> ***The physician must be the main source of information for patients taking medication and should have available PPIs to discuss with the patients.
> ***The physician should have the final authority to withhold patient information when he or she judges that it could hamper or further complicate the treatment process or believes the patient's reaction would be detrimental to his or her physical or mental health.[6]

The nature of the paternalistic model itself—and the examples cited—should be enough to make it clear that, in such a context, the concept of a right to refuse treatment is utterly senseless. The images of infancy and insanity here overshadow all other considerations. The subject is considered to be *like* an infant or an insane person. Since it is further assumed that the physician is not only caring, but also competent, it follows that the treatment the physician chooses for the patient is the treatment that patient would want were he or she old enough or mentally healthy enough to exercise proper judgment. Although it makes sense to reject this model, it makes no sense to retain the model and still insist on the patient's "right to refuse treatment." Yet this is precisely what all the present-day mental-health lobbyists do who accept the proposition that some persons are "mentally sick" enough to be committed, but nevertheless possess a "right to refuse treatment."

III

I shall use the term "communism" here to designate all types of collectivistic modes of health care. The communist model resembles the paternalistic one, in that the physician is not the patient's agent, but is an agent of a system of political values or of "social welfare."

Of course, in a collectivistic perspective, the doctor is still regarded as striving to make his sick patients healthy—but sickness and health acquire an increasingly political coloration. Also, the physician is viewed as an agent not only of the patient, but also of the patient's family, employer, and the state itself. Since such a perspective requires the denial of conflicts of interest between these parties (which may generate different definitions of illness, health, and treatment), raising the question of whose interests the physician serves is, in this context, professional heresy. The communist physician works for health, which is in everyone's best interest. Likewise, the democratic psychiatrist works for mental health, which is also in everyone's best interest. Such job definitions are ambiguous and evasive, but these characteristics are this model's greatest strengths.

It would be foolish to pretend that most people most of the time want to think clearly and act conscientiously about the vexing moral dilemmas of their daily lives. Moreover, in its simplest form—exemplified by public health and sanitary policies—the collectivistic model of health care can easily appear to satisfy everyone's best interests. Who can be against potable drinking water and sewage disposal? Ibsen has answered this question with the characteristic prescience of the artist in *The Enemy of the People.*

Still, this sanitary model of health care is exceedingly popular— partly because the vast majority of people in modern societies agree that the state should engage in measures to prevent the spread of, say, amebiasis or malaria. However, this should not blind us to the fact that behind such policies lies a social consensus that is, in principle, not very different from the social consensus of earlier societies concerning policies aimed at maintaining religious purity. The idea that it is good to be and to stay healthy is, in fact, like the idea that it is good to be and stay religious: both are values that many people reject some of the time and some people reject all of the time. The values of health and longevity constantly compete with a host of other values —such a pleasure from eating and smoking, cultural dietary habits, patriotic duty, and so forth.

Since the collectivistic model of health care rests squarely on social consensus (and the power of the modern state to mold and enforce it), it is incompatible with a concept of a right to refuse treatment. If every individual's good health is an asset to the group, and every individual's ill health is not only a burden but a danger to it, a right to refuse treatment is tantamount to a right to endanger and injure the other members of the community. In such a perspective, there are, strictly speaking, no individual rights, there are only rights adhering to persons as members of the group: The individual is accepted as a person only if he or she abides by the medical precepts of the group; if not, the individial is expelled—as insane (or, more rarely, as a criminal). The saga of Wilhelm Reich and his orgone box provides a dramatic illustration of this principle. And so does the example presented by the current conventional medical-legal posture toward the person who threatens suicide.

What happens when the person tells the doctor that life is not worth living and that he or she is going to commit suicide? Usually, and especially if the doctor is a psychiatrist, that doctor becomes very upset. The doctor reacts much as a priest might react to a person who desecrates the crucifix. If someone plans to kill herself or himself, that person should not tell it to a psychiatrist unless prepared to face the highly predictable consequences of the action. It may be objected that a psychiatrist need not react with punitive sanctions (called "treatment") to the "suicidal patient"; that the physician could eschew such a response and accord the individual the right to determine his or her own fate, including the choice of suicide. Of course a psychiatrist could do that, but not within the model of collectivistic health care.

IV

Finally, we come to the capitalist or contractual model of medical care—a perspective just as familiar as, if not more so than, the two models previously discussed. In this view, medical care is a service, more or less like any other. In the United States, the dispensing of some medical services remains in this contractual model. People buy and sell eyeglasses, dental work, face lifts. These are contractual arrangements, one party selling a service and another buying it. In this context, it is foolish—because it is trivially true—to speak of a right to refuse treatment. We speak of a "free choice" of medical care, instead.

This approach to medical care is simple, appealing (at least to libertarians), but troublesome. Why? Because it applies only to persons who can make contracts—which immediately rules out children. It also presents problems with respect to poor people—who, of course, cannot contract to buy many medical services, just as they cannot contract to buy many other things. (Indeed, it is because they cannot afford many things that they are called "poor.") Hence, most people now reject the contractual model of medical care—feeling (in part, rightly) that a poor person may "need" penicillin or an appendectomy in a quite different way than such a person could be said to "need" a Cadillac or a Caribbean vacation.

Like children, the insane too have traditionally been considered to be incapable of making valid contracts. The needs of such persons and the complex moral and social problems that their care and welfare pose furnish additional reasons for rejecting the capitalist model of health care. This rejection has gone much farther than many people realize. It is worth mentioning in this connection that, although the United States is still considered to be a capitalist country, we possess a commodity in more than ample supply that we nevertheless cannot sell or buy without extremely stringent restrictions. That commodity is a hospital room. It does not matter how much money a person might have, he or she cannot buy a stay in a hospital longer than is ostensibly required by his or her condition. Suppose that a person goes to a hospital for the repair of a hernia or an appendectomy and is considered to be medically ready to be discharged in four or five days. Suppose the patient then says to the doctor (or the administrator of the hospital), "I like it here. I want to stay two more weeks." The patient would not be allowed to stay—even if the hospital were half empty and even if the patient were willing to pay twice the rate normally charged for a room. A person can no longer do this in America—but can still do it in Switzerland.

Having remarked on the disadvantages of the capitalist model of health care, let us now note its advantages, which are no less impressive. Most importantly, only in such a contractual arrangement does the patient retain control of the medical relationship, and thus the power to protect herself or himself from physicians (and the state, insofar as the state controls medical care). Patients hire the doctors they want and fire them if they are dissatisfied with the service. Physicians, for their part, also benefit—mainly by being spared the conflict of divided loyalties. They contract to be their patients' agents and no one else has any claims to intrude into that relationship. Adherence to such a model would eliminate most of our present-day problems of malpractice, bed utilization, access to patient records, drug abuse, and so forth. The implications of such a model—for individual liberty and the quality of medical care on the one hand, and for regulating and distributing such care on the other—are, of course, far-reaching, but need not further concern us here.

V

I have argued that the concept of the "right to refuse treatment" —especially in its current usage as the hospitalized mental patient's right to refuse psychiatric treatment—is either deceptive or nonsensical. My reasons for this judgment may be summarized as follows:

1. In the paternalistic model of medical care, the relationship between doctor and patient is defined as a fiduciary one. This makes the patient unable (or less able than the physician) to evaluate the treatment the doctor deems necessary for his or her illness; and it makes the treatment a beneficial intervention which patients, were they able to judge its value properly, would themselves eagerly seek and willingly receive. Thus construed, treatment is, *ipso facto*, a "good" that it makes no sense to reject. Since only the ignorant and the insane would reject such treatment, and since the model of paternalism explicitly deprives them of the right to self-care, it is oxymoronic to assert that they have a "right to refuse treatment."

2. In the communist or collectivistic model of medical care, health is as much a concern of the community as it is of the individual. The doctor, as the protector and promoter of health is, therefore, the agent of both the individual and the community (state). In this model the individual has no rights against those of the community or collectivity: the patient's health is a benefit to the group and any illness is a detriment to it. In this view, there can be no "right to be sick," since such a right would entail a direct or indirect aggression against the group. Hence, there can also be no "right to refuse treatment," since it, too, would constitute an aggression against the group.

3. In the capitalist model of medical care, medical treatment is, ideally, a service, like repairing shoes or teaching dancing. It is,

of course, true—but it is so obvious that it would be stupid to assert it—that golfers have a right to refuse golf lessons or that housewives have a right to refuse to buy lettuce. Likewise, it is true that, in this model, the patient has a right to refuse treatment —but this truth is both trivial and tautological. Since medical treatment is a contract for a service that, ideally, both doctor and patient enter into, and exit from, freely, it is redundant to assert that the patient has a right to refuse treatment (or that the doctor has a right to refuse to treat the patient—or, indeed, to accept the client as a patient).

The "right to refuse treatment" is thus either a trivial truth, without real meaning—or a pretentious falsehood, concealing certain unpleasant facts about the real medical (psychiatric) world. Why, then, has the phrase "the right to refuse treatment" come into being and why is it so popular? What are its real — economic, political, and social—uses and consequences?

The answers to these questions lie beyond the scope of this presentation. Let it suffice to say here that the concept of the "right to refuse treatment" serves many of the same functions that are served by the other key terms of modern psychiatry—such as mental illness, diminished capacity, psychiatric hospitalization, and so on (some of which are losing their credibility and hence their utility). And what is that purpose? It is, as I said about "mental illness" more than twenty years ago, "to disguise and thus render more platable the bitter pill of moral conflicts in human relations."[7]

More narrowly and specifically, however, the concept of a right to refuse treatment presupposes and yet obscures the fact that it applies only to social contexts in which "treatment" is compulsory. In other words, it presupposes and yet obscures the existence of laws and court orders forcing certain people to submit to certain procedures deemed to be therapeutic. Although mental patients are the main "beneficiaries" of such involuntary treatments, they are not the only ones. In some jurisdictions, prostitutes are compelled by law to submit to injections of antibiotics.[8] In at least one case, the courts have upheld the decision of prison authorities to impose kidney dialysis on "a mentally competent prison inmate" suffering from end-stage renal disease who had refused hemodialysis.[9] Those who want to abolish such "therapeutic" coercions, should seek the remedy in the repeal of legislation permitting or mandating compulsory medical or psychiatric interventions—not in a fictitious "right to refuse treatment."

Acknowledgment

Adapted from a lecture delivered at the Conference on Health Care Ethics, University of South Carolina, Columbia, South Carolina, November 9, 1979.

References

1. Szasz, T. S. The Right to Health, *The Georgetown Law Journal* **57**: 734–751 (March, 1969); reprinted in *The Theology of Medicine.* New York: Harper/Colophon, 1977, pp. 100-117; and *Psychiatric Slavery: When Confinement and Coercion Masquerade as Cure.* New York: The Free Press, 1977.
2. Szasz, *Psychiatric Slavery*, op. cit., especially Chapters 1 and 8.
3. See, for example, Sadoff, R. L., On Refusing Treatment: Rights and Remedies, *Advocacy Now* **1**: 57–60 (August, 1979); Schaeffer, P. Court Rules That Mental Patients Have Right to Refuse Treatment, *Clinical Psychiatry News* **8**: 1 & 36-37 (January, 1980).
4. Ref. #1.
5. Cited in Westin, A. F. Medical Records: Should Patients Have Access? *Hastings Center Report*, December, 1977, pp. 23–28; p. 23.
6. Cited in, APA objects to proposed drug inserts, *Psychiatric News* **14**: 16 (December 7, 1979).
7. Szasz, T. S. The Myth of Mental Illness. *The American Psychologist* **15**: 113–118 (February, 1960), p. 118; reprinted in *Ideology and Insanity*, pp. 12–24. Garden City, N. Y.: Doubleday Anchor, 1970, p. 24.
8. Slovenko, R., Prostitution and STD: When Treatment Becomes Punishment. *Sexual Medicine Today*, January, 1980, pp. 14–15.
9. Involuntary care approved, *American Medical News*, December 21, 1979, p. 14.

Chapter 9

Refusal of Psychiatric Treatment

Autonomy, Competence, and Paternalism

Ruth Macklin

I

Let me begin by quoting from a short news article that appeared in the *New York Times* on November 1, 1979. Entitled "Judge Curbs Forced Medication in Treatment of Mental Patients," the article raises a number of central issues that receive a detailed analysis below:

> Federal District Judge Joseph L. Tauro has ruled that a mental patient has a right to refuse tranquilizing medication unless he is likely to become violent. The judge called such forced medication "an affront" to human dignity.
>
> Judge Tauro ruled Monday in favor of former patients at Boston State Hospital who had filed a class action suit seeking restrictions on the use of mind-altering drugs and on seclusion at the facility.
>
> "Given a nonemergency, it is an unreasonable invasion of privacy and an affront to basic concepts of human dignity to permit forced injection of a mind-altering drug," Judge Tauro ruled...

"The desire to help the patient is a laudable if not noble goal," he said. "But a basic premise of the right to privacy is the freedom to decide whether we want to be helped or whether we want to be left alone."

He rejected a contention by attorneys for the hospital that a patient, whether voluntarily or involuntarily admitted, was incompetent to decide whether to accept treatment.

"The weight of the evidence persuades this Court that, although committed mental patients do suffer at least some impairment of their relationship to reality, most are able to appreciate the benefits, risks and discomfort that may reasonably be expected from receiving psychotropic medication," he said.

Forced injection of mind-altering drugs also violates the provisions of the freedom of speech guarantees of the First Amendment, he added.

"The First Amendment protects the communication of ideas," the judge said. "That protected right of communication presupposes a capacity to produce ideas. As a practical matter, therefore, the power to produce ideas is fundamental to our cherished right to communicate and is entitled to comparable constitutional protection.

"Whatever the powers the Constitution has granted our Government, involuntary mind control is not one of them, absent extraordinary circumstances."

Although this recent ruling reported in the newspaper addresses the issues from a legal standpoint, rather than from the perspective of moral philosophy, the concepts and principles involved are virtually identical.

Before turning to the specific issues raised by the judge's ruling —including issues that arise from the way he chose to formulate his decision—I want to lay down several premises on which my later arguments will be based. First of all, it is too narrow an approach to look at the question of refusal of psychiatric treatment as if the only value at stake were the freedom or liberty of the psychiatric patient. Decisions involving refusal of treatment are often couched in terms of the patient's "right to decide." Although I think this *is* the central issue in cases in which the patient is competent, other considerations enter into the picture when the patient is either clearly incompetent or when there is uncertainty surrounding the patient's competency. In fact, it is just this question of competency that Judge Tauro's decision glosses over much too quickly—a question to which I shall return. My first premise, then, is to acknowledge that freedom is a fundamental value—one that should be respected and promoted in medical (psychiatric) settings, as well as elsewhere in personal and social life.

Following quickly upon this first premise is a second: though freedom is a value of fundamental importance, it is not the only value that needs to be preserved, promoted, or respected in medical practice. To name just a few other salient values, we have justice, equity, equality, self-respect, self-esteem, autonomy, preservation of life itself, improving the quality of life, benevolence, and humaneness. Some of these values are more typical in medical contexts than others and, as we are all too often reminded, these cherished values often come into conflict with one another, making it impossible to satisfy more than one simultaneously.

My third premise is this. Freedom to choose is a *hollow* value if the individual facing a choice lacks the capacity or the opportunity for rational deliberation, adequate understanding, or reasonable assessment of the consequences of the options being offered. The discerning reader will, of course, recognize all the baggage I have packed into this third premise. What constitutes the *capacity* for rational deliberation? Or *adequate* understanding? Or *reasonable* assessment of consequences? What do we mean by "rational," "reasonable" and "adequate"? I do not intend to dodge these hard questions entirely, and will come back to them shortly. But however difficult it may be to arrive at a precise formulation of the meanings of these problematic notions, and however hard it may be to devise criteria for their correct application in practice, these difficulties should not lead us to dismiss the premise entirely. Sometimes hard questions are just that—hard. They are not resolved by pretending that conflicting values do not exist, or that people are less complex than they really are. Now I realize that it is not very illuminating to point out that an incompetent patient is one who lacks the capacity for rational deliberation, adequate understanding, or reasonable assessment of the consequences of the options that are available. But neither is it helpful to deny that the distinction between the competent and the incompetent patient is relevant to the question of the patient's right to refuse psychiatric treatment.

What values are central to debates surrounding the right to refuse psychiatric treatment? I have already mentioned individual liberty or freedom. The term now widely used to denote *limitations* on liberty, justified by appeal to a person's own good, well-being, happiness, or interests, is "paternalism." Paternalism has enjoyed an especially bad press in recent years, particularly in areas related to medical treatment, the doctor–patient relationship, and the actions of the FDA and other government agencies that regulate the actions of citizens, presumably for their own good. But are all paternalistic acts and practices unjustified? I shall argue shortly that although many are unreasonable, there still remains a class of justifiable paternalistic acts. These are ones in which the person or persons being coerced are incompetent. But before mounting that argument, let me introduce another key concept I will need for developing my further remarks about refusal of psychiatric treatment. The third major concept to be invoked in what follows is that of *autonomy*. To be autonomous is to have a "self-legislating will," in Kantian terms. It is to be author of one's own beliefs, desires, and actions. The autonomous agent is one who is self-directed, rather than one who obeys the commands of others. These various senses of autonomy presuppose an idea of an authentic self, a self that can be distinguished from the reigning influences of other persons and, more importantly for our present topic, a self that has a continuity of traits over time. I will argue below that if individual autonomy is a value that should be protected, promoted, and preserved, libertarians should recognize that autonomy may be lost by a patient's refusal of psychiatric treatment— treatment that might have prevented deterioration, humiliation, or decline. But that is a conclusion, rather than a premise of my argument, so let me turn next to some philosophical considerations— conceptual as well as moral.

II

It is clear that coercion for paternalistic reasons limits people's freedom. Even if arguments in favor of controlling the behavior of rational adults for their own good remain unconvincing, there is still a question about those who are less than fully competent.

The principle of liberty, as urged by John Stuart Mill and others, appears to prohibit paternalistic interferences with anyone's liberty of action for reasons other than that of protecting innocent people from harm. It seems to rule out intervention into the behavior of suicide attempters, electric shock treatments for mental patients who are opposed to them, and involuntary commitment of those who are judged mentally ill and in need of care, custody, or treatment. But is this the proper way to interpret Mill's writings?

The word "paternalism" derives its meaning from the notion of treating others in what the dictionary describes as a "fatherly" manner. The clearest cases of paternalistic acts are found in the behavior of responsible parents or caretakers toward young children. Not only do we think we should interfere with the liberty of action of children in order to protect them from harm, we believe it is our duty to coerce or limit small children in many ways. There is usually no quarrel, in principle, with this view. Debates and quibbles begin over the particular age at which intervention ceases to be warranted or about just which forms of coercion are allowable, and in what areas of a child's life. But it is generally held to be fully justifiable to limit children's liberty of action both in order to prevent them from destroying themselves and also to foster their growth and thriving. At least in the case of very young children and probably even up to adolescence, paternalistic reasons for controlling behavior appear sound.

The clearest cases of unjustifiable paternalistic acts are those in which obviously rational adults, who know what they are doing, are coerced against their wishes presumably for their own good. This is surely the group Mill had in mind when he asserted his libertarian principle, as a closer look at his position reveals. Mill would no doubt support a justification for involuntary commitment where a high probability exists that a person judged dangerous to others will commit an act of violence or other harm. It hardly seems as though Mill could support involuntary commitment on the grounds that someone is "dangerous to self," for that would be paternalistic. But would it be an unjustifiable form of paternalism? Caution is needed to avoid treating every paternalistic act as wrongful interference, even on a view that appears at first glance as strongly anti-paternalistic as Mill's. In applying his principle, we need to ask: does the prohibition against interfering with another's liberty apply to *everyone's* liberty? Does Mill himself recognize any exceptions to the prohibition against paternalistic acts? Mill's answer is explicit, but not wholly clear:

> It is, perhaps, hardly necessary to say that this doctrine is meant to apply only to human beings in the maturity of their faculties. We are not speaking of children, or of young persons below the age which the law may fix as that of manhood or womanhood. Those who are still in a state to require being taken care of by others, must be protected against their own actions as well as against external injury. For the same reason, we may leave out of consideration those backward states of society in which the race itself may be considered as in its nonage.... Despotism is a legitimate form of government in dealing with barbarians, provided the end be their improvement, and the means justified by actually effecting that end. Liberty, as a principle, has no application to any state of things anterior to the time when mankind have become capable of being improved by free and equal discussion.[1]

Even though Mill's focus lay on the laws and actions of government, rather than on acts or practices of psychiatrists, the issues in both contexts are largely the same. The major task is to produce adequate criteria for judging when individuals are "in a state to require being taken care of by others" or when they are not (yet) "capable of being improved by free and equal discussion." In spite of his stalwart defense of the libertarian principle, Mill might well be prepared to accept a wide range of paternalistic interventions— precisely those directed at human beings who have not attained "the maturity of their faculties." He offers no clear or practically workable criterion for distinguishing between those who have attained such maturity and those who have not; but he nonetheless countenances a class of cases of justified paternalism. Not only does Mill approve of the imposition of paternalism in the case of children—those who are "still in a state to require being taken care of by others." He actually urges a paternalistic line of conduct. He is quite explicit in holding children and "those backward states of society in which the race itself may be considered as in its nonage" exempt from his injunctions against paternalism. Notice also that Mill did not need a notion of "mental illness," or even that of competency, to refer to the condition he describes here.

But what about other individuals who are "child-like" in a number of ways, especially in their inability to be "improved by free and equal discussion"? Among these are a majority of those correctly labeled mentally retarded, many who suffer some form of mental illness, emotional disturbance or behavior disorder, and the senile. Members of all these groups manifest lack of full-scale rationality as a permanent or relatively enduring trait. Injunctions against paternalism seem not to apply in cases where people satisfy Mill's rather vaguely worded conditions—"being in a state to require being taken care of by others" and "incapable of being improved by free and equal discussion." Recall that Mill, in applying his own principle, claims that "despotism is a legitimate mode of government in dealing with barbarians, provided the end be their own improvement." This brand of paternalism is morally and politically unacceptable in today's world climate of anticolonialism. Yet it is consistent with this view to hold that coercion of people who are clearly nonrational or incompetent to manage their own affairs is a legitimate mode of treatment, provided the end be their own improvement. Mill says his doctrine is meant to apply only to human beings "in the maturity of their faculties." On one plausible reading of this condition, retarded persons, the senile, and those afflicted with what psychiatrists call "thought disorders" are not in the maturity of their faculties in any but the chronological sense of "maturity." The relevant sense of "maturity" here is having "fully developed" mental capacities, where the lack of full development may be a result of chronological immaturity (children), decline of mental faculties (the senile), disease (mental patients), or unexplained developmental failures (many mentally retarded persons). The law now recognizes an intermediate category termed "diminished capacity"—a category used to mitigate criminal responsibility for some persons who have committed offenses.

So even an attack on paternalistic interference as strong as Mill's leaves room for a measure of justifiable paternalism in controlling, modifying, or improving the behavior of less than fully rational individuals. For those judged rational (in some appropriate sense of that complex and slippery notion), paternalistic intervention would not be justified. Since the profoundly retarded and the severely depressed have many of the same characteristics as helpless children, the same moral principle of humaneness applies: do not let harm befall those who, if left alone, would probably perish. In the case of parents and children this moral precept is strengthened by the role parents occupy and the special duties and obligations that flow from that role. The borderline cases pose the most difficulty, since the arguments could go either way depending on how close to the paradigm the analogous cases are drawn. If the notion of being inhumane has any moral force at all, it applies to situations where those who are clearly psychologically incapable of caring for their own needs are simply left to fend for themselves or perish.

A major problem lies in the fact that disagreement exists over where to draw the line conceptually between competence and incompetence. Part of the reason for this disagreement is that competence and incompetence are not discrete, separable states. One shades into the other, and people may be highly competent in some respects yet wholly incompetent in others. Failure to recognize this can result in mistreatment of those who are classed as incompetent according to specific criteria recognized by the law. For example, those judged incompetent to manage their own finances or to make a will may still be sufficiently rational to grant or to refuse consent for treatment using some medical procedure—say, electroconvulsive therapy. To be incompetent in some ways is not necessarily to be incompetent in all ways. Some mental patients suffer primarily from mood disorders, yet their cognitive capacities remain basically intact. The mildly retarded are slow to grasp concepts and may be incapable of abstract reasoning, yet they may understand, when it is carefully explained, what is involved in sterilization. All this suggests the need for a notion of variable competence, which would allow for a cluster of criteria to be used when determinations of competency must be made. This would result in appropriately specific judgments of competence rather than the kind of global assessment that is either too sweeping or turns out to be false. In the end, however, disagreement may remain over the question of whether people ought to assume responsibility for others (aside from children and maybe adult relatives) who cannot care for themselves.

III

The problem now shifts to the search for appropriate criteria for making judgments of rationality or competence. It is usually easier to get clear about things by looking at paradigms, or clear cases, than by offering definitions or criteria, however. Normal adults are presumed to be rational or competent, and so are allowed to go about their business without interference. Five-year olds are generally presumed incompetent in many respects, and so are watched carefully and their behavior interfered with, when the need arises, by caretaking adults. With the retarded and the mentally ill, it is simply unclear where the general presumption ought to lie. We seek to avoid erring in either of two opposite directions: being too paternalistic, and therefore unjustifiably coercive; or too permissive, thereby opening the door to self-destructive or other irresponsible acts. The one evil consists of violating the cherished value of individual freedom; the other, allowing harm, destruction, or even death to befall an innocent, helpless human.

It is dismaying, but not too surprising, to discover that the grounds on which judgments of competency are made—usually by psychiatrists, in courts of law—are neither wholly objective nor generally agreed upon by experts in the mental health field. There just is no generally accepted, overall *theory* in psychology or psychiatry on which to base judgments about an individual's rationality, competence, or sanity. Instead, there exist a variety of sub-theories and isolated criteria, a small number of which psychiatrists adhere to when they are legally authorized to testify in court.

Yet even in the absence of a solid theoretical foundation for making legal judgments about competence—judgments that may have direct consequences for licensing psychiatric treatment—wide areas remain in which people's competence needs to be assessed. The one that concerns us here is the need to gain informed consent from patients for treatments of various sorts. There is not a great deal of helpful written material on the meaning of "competency" in this context, but a recent article in a psychiatric journal advances the discussion considerably. In an article entitled "Tests of Competency to Consent to Treatment,"[2] the authors describe the various tests of competency used today. They cite (1) evidencing a choice; (2) "reasonable" outcome of choice: (3) choice based on "rational" reasons; (4) ability to understand; and (5) actual understanding, as the five basic categories proposed in the literature or readily inferable from judicial commentary. The authors note further that a useful test for competency is one that can be reliably applied, is mutually acceptable or at least comprehensible to physicians, lawyers, and judges, and finally, is capable of striking an acceptable balance between preserving individual autonomy and providing needed medical care.[3] Thus, although a firm theoretical basis for making judgments of competency may still be lacking, efforts are underway to develop sound practical criteria that psychiatrists can apply objectively.

If there are, as I would urge, justifiable as well as unjustifiable instances of paternalism, then it is a mistake to deplore all paternalistic interventions as unethical because they interfere with an individual's freedom. Individual liberty can only be exercised when a person is in a reasonable state of health (mental or physical) and has reached an adequate age or degree of competence. Unless we hold—as some people do—that liberty is more important than life itself, limitations on freedom are justified in the interest of preserving life, health, and even autonomy.

The general issue of paternalism and its possible justifications should, moreover, be considered in connection with who serves as the intervening agent: outside authorities, such as the state; parents or relatives; or caretakers from private or voluntary groups. Some responsibilities and authorities derive from specific roles, such as those of teacher, physician, employer. In some cases, the right to intervene in various ways is contractually determined. In other cases, authority to act is an implicit yet integral feature of an established relationship. The need to make these distinctions becomes evident upon the realization that not all paternalistic interventions should be seen as unjustifiable simply because they involve a measure of coercion. The problem is compounded by the fact that invasive procedures and forcible means can be used to attain highly desirable ends.

Take, for example, compulsory drug treatment or electrical brain stimulation performed on those who engage in acts of self-mutilation; or behavior modification using aversive conditioning on the mentally retarded or on autistic children; or medication to subdue those suffering from what psychiatrists call "manic flight." Some patients after treatment are more self-reliant, function more independently, and enjoy a heightened sense of well-being. These qualities are, by general agreement, judged to be desirable or positive ones, and should therefore be considered objective criteria for the purpose of determining what "really is" in a person's best interest. In these situations, the subjects' *dignity* is enhanced—the quality Judge Tauro invoked in his decision; they are enabled to function more autonomously; their self-satisfaction is increased; they are helped to thrive, rather than merely to subsist. These are reasonably clear cases. But what about paternalistically justified modes of behavior control used in prisons and social control of deviants as practiced in mental institutions? Are these actions so obviously for the subjects' own benefit? They are surely practices that those in control, given their commitment to predominant social values and institutional norms, deem to be for the good of those controlled. If the subjects of control disagree about what is in their best interest, their failure to concur may be taken by those in power as evidence that they are less than fully rational and so may justifiably be coerced for their own good.

Consider now how a range of facts about the special populations on whom paternalistic regulation is practiced aid in drawing moral conclusions. The use of powerful behavior control techniques with severely retarded or disturbed persons can often result in their functioning more independently or with greater self-reliance, thus enhancing their own well-being as well as making them less dependent on others. Using the same powerful technologies on prisoners may, in contrast, reduce their autonomy, render them more passive and, perhaps, more submissive to the will of others. It is hard to arrive at a fully satisfactory, objective account of which characteristics constitute "changes for the better" as a result of psychiatric treatment. In accordance with the Kantian precept that dictates respect for the dignity of human beings, it is incumbent on theorists and practitioners of behavior control to promote the welfare of all persons, even those who are functionally incapacitated or deemed socially undesirable. In the absence of universally agreed upon criteria for determining what really is in the best interests of the subject, paternalistically justified modes of behavior control directed at rational or even incompetent persons should be employed only with great caution.

In the case of the mentally retarded, their level of competency can actually increase as a result of special education, vocational training, behavior modification, or simply undergoing life's experiences. Nevertheless, their retardation will almost surely remain an enduring trait rather than a temporary condition. So the issue of paternalistic intervention may continue to arise throughout their lives, with little likelihood of their attaining a full state of normalcy.

The situation is somewhat more complicated with the mentally ill, however, since there is greater likelihood that their condition is a temporary one. Consider, for example, a mental patient who has intermittent lapses and remissions and who begs—while lucid—not to have electric shock treatment administered. Is ECT, aimed at improving the patient's condition, justified during lapses when he or she may correctly be held incompetent or irrational? Or should the wishes of such patients be honored on the grounds that their desires were expressed at a time when they were rational agents—persons who deserve not to be treated paternalistically?

Perhaps the hardest cases are those of so-called "manic flights." These persons appear rational by virtue of their verbal coherence and occasional bursts of artistic and other creativity. What's more, people who undergo these episodes often refuse even to see a psychiatrist, much less submit to treatment. At least part of the reason is that they enjoy the hypermanic experience.

So why is intervention of any sort felt to be necessary or even desirable? A colleague of mine who is a psychiatrist described some cases to me. In some intances, those who underwent manic episodes engaged in public behavior that threatened their careers, their reputation, their family's well-being, and perhaps even their own self-respect. Sexual exhibitionism, giving away all their money and property, soliciting 12-year old boys, often leads such persons to suicide attempts or other self-destructive acts. In most cases, those who were victims of hypermania expressed overwhelming gratitude—once they returned to "normalcy"—to psychiatrists, family members, and others who had intervened, even over their protests. In all cases, those who exhibit such behavior act uncharacteristically. It is not simply that they act differently from other people—that they are "deviant," in the usual sense. They are deviant in the additional sense that they depart from their *own* established character. They are not their *true selves*, they lack continuity with their typical or normal or characteristic personality. Such people could be said to be wholly lacking in autonomy when in those states. If they are offered psychiatric treatment and refuse it, if they are cajoled or manipulated or even coerced into therapy of some sort, is their autonomy being violated? I think something more like the opposite is true. All evidence points to a lack of genuine autonomy in those who behave out of character in hypermanic, self-destructive ways.

I have tried to avoid as much as possible using terms like "mental illness," "emotional or psychiatric disorder," "sanity and insanity," and other concepts that have aroused so much debate in scholarly and now popular literature. This is because I think none of the issues I have raised or the arguments I have examined require that we settle those debates or even that we come down on one side or the other. Nothing whatever hangs on whether mental illness is a myth, whether personality disorders are diseases, whether there should be a legal insanity defense, and so on. I have been talking only about the concepts of competence, paternalism, and autonomy and their applications in the context of refusal of psychiatric treatment. I surely have not given an adequate analysis of those concepts here, nor have I been very thorough in describing or analyzing the examples I chose to illustrate them. Yet there remain solid moral principles—principles of humaneness, of benevolence, of concern for the interests and well-being of others who demonstrably are not competent to act in their own interests. *Autonomous* agents who act in ways that others believe are against their own interest have the right to be let alone. The truly incompetent lack full autonomy, and so that quality cannot be violated by imposing treatment. If there is a reasonable likelihood that their autonomy can be preserved or restored by medication, then forced treatment of such patients is warranted. I am, however, opposed to more invasive treatments being imposed involuntarily, such as psychosurgery or ECT, for purely *behavioral* symptoms.

There will probably always remain some degree of conceptual uncertainty surrounding the notions of freedom, competence, and autonomy. And there will surely never be agreement on which moral principles, or which values ought to prevail when they come into conflict. Yet a careful, systematic approach will yield solutions that are morally superior to those arrived at by a dogmatic adherence to a stance that either rules out paternalism in principle, or else accepts its legitimacy uncritically.

Acknowledgment

Portions of this article appear in an earlier form in Chapter 3 of *Man, Mind, and Morality:* The Ethics of Behavior Control, Englewood Cliffs, New Jersey, Prentice–Hall, 1982, by Ruth Macklin.

References

1. Mill, John Stuart. *On Liberty*, in Max Lerner, ed., *Essential Works of John Stuart Mill*. New York: Bantam Books, 1961, pp. 263–264.
2. Roth, Loren H., Meisel, Alan, and Lidz, Charles W. "Tests of Competency to Consent to Treatment," *American Journal of Psychiatry* **134**: 3 (March 1977).
3. *Ibid.*, p. 280.

Chapter 10

Consent and Competence
A Commentary on Szasz and Macklin

James L. Stiver

In his "Dissenting View," Thomas Szasz makes two main points about the right to refuse treatment: it "makes no sense" to say that psychiatric patients have such a right, and it is a deceptive thing to say.

In support of the first claim, Szasz examines three currently prevalent medical models—paternalistic, communistic, and capitalistic—and considers what it would mean to say, in each model, that psychiatric patients have a right to refuse treatment. He argues that the paternalistic model assumes that the physician is not only caring but competent, whereas the patient is not expert about what would be good for him or her—*especially* when a mental patient. Hence the physician should have the final authority about treatment, and the right to refuse treatment makes no sense on the model. In the communist model, there is no assumption that the physician's purpose is to enhance the well-being of the patient as an end in itself. Rather, the physician is the agent of others—ultimately, the state—and therefore this physician's primary purposes are political, not medical. The right to refuse treatment cannot be accorded to the patient in this model, for to do so would be to grant the patient the right to injure the community by rejecting its values. So here again the right to refuse treatment makes no sense.

We reach the same conclusion in the capitalist model of medical care, but for very different reasons. When medical care is a service to be contracted for, *of course* the patient has the right to refuse treatment, but this is obvious, so "trivial," that it is "foolish" to speak of it, and "stupid" to assert it.

It seems to me that the point Szasz wishes to make could be made better by avoiding the rubric "makes no sense." One suspects a Wittgensteinian orientation; but Szasz could sidestep methodological controversies by speaking instead of incompatibilities and logical consequences—as indeed he does, though not uniformly. Thus, cast in these terms, Szasz has argued that the right to refuse treatment is incompatible with the paternalistic and communistic models, but is part and parcel of the capitalistic model.

Not only would this way of putting the point allow Szasz to skirt methodological issues, but it would also avoid creating the appearance that Szasz is somehow opposed to patients having the right to refuse treatment. His libertarian orientation must commit him to such a notion (at least for competent adults, and perhaps for some others), but how can he endorse this view if he also holds that the concept makes no sense? Szasz ought to make the point that whether patients have the right to refuse treatment is model-relative without making it in such a way as to foreclose his endorsing such a right (via endorsing a model within which it is recognized). In short, unless I misunderstand Szasz's position, the way he puts his point tends to be misleading. This is most unfortunate, because it is a very important point to make, particularly in the context of psychiatry.

In any event, to the extent that psychiatry does not invoke the capitalist model—and involuntary psychiatry surely does not—it is at best deceptive for psychiatrists to speak of the right to refuse treatment. Only those patients regarded by psychiatrists as competent have such a right, and they have no need to exercise it, not being subjected to treatment they do not want. Those judged to be incompetent are subjected to such treatment, but it is for their own good (or so it is claimed, at least), and they do not want it only because they are incompetent.

Ruth Macklin's paper provides a much needed examination of the concept of competence and its relations to liberty, autonomy, and paternalism. Where Szasz places priority on the value of liberty, Macklin cites other cherished values that often conflict with liberty, e.g., autonomy (frequently confused with liberty), humaneness, life itself, and others. In particular, she is concerned to argue that despite the strong claims of liberty, there are some actions in a psychiatric context that are paternalistic, yet justified. Without going into the details of her conceptual analyses, those actions may be characterized as follows: the patient is incompetent and it is a reasonable likelihood that the treatment will preserve or restore autonomy. Her reasoning is that if the patient is truly incompetent, full autonomy is lacking. But one who lacks autonomy—one who is incapable of self-legislation—is not really free to choose in anything but a hollow sense. To allow that person to exercise a right to refuse treatment is to invite self-inflicted harm in a most profound way: by increasing the probability that at some future time the patient will be less autonomous than at the time of the refusal of treatment. Thus, the liberty of incompetent psychiatric patients may be violated justifiably (presumably in direct proportion to their degree of incompetence).

The rub in this impressive if not wholly compelling argument is the concept of competence. Macklin states repeatedly that this is a difficult concept to pin down, but even when she is most circumspect, she underestimates the difficulties. She raises the hope of finding practical criteria of competency "that psychiatrists can apply objectively." Although no doubt these criteria will employ value concepts —ultimately dealing with what is good for the patient—they will be objective to the extent that they are "by general agreement, judged to be desirable or positive" values. But it is precisely in those cases in which the psychiatrist and the patient disagree about what is desirable for the patient that the patient is likely to be judged incompetent. It seems that the patient's disagreement is not to mitigate against the "general agreement" because he or she is incompetent. And, of course, the patient is incompetent because he or she is not party to the "general agreement."

Does this mean that a person is to be regarded as incompetent merely because that patient does not share the psychiatrist's values? Or even any psychiatrists' values? Suppose a person were to be placed in the power of another who proposes to subject him to exquisite tortures that will last indefinitely, but will someday cease, after which time the person will be free and improved (in terms of that person's *own* values). Suppose further that the subject is given a capsule and told that if broken it will insure instant, painless death. Suppose finally that the person does indeed attempt to commit suicide. (The capsule is in truth ineffective.) Should the patient now be judged incompetent, even if most other people do not place such a high value on liberty or freedom from pain as that person does? Why can it not be rational for such a patient to decide that it is just not worth it, in terms of that patient's own values? Macklin urges libertarians to "recognize that autonomy may be lost by a patient's refusal of psychiatric treatment." Yes, but by the same token, autonomy may be realized by a patient's refusal to judge his or her welfare in terms of the psychiatrist's values. Autonomy is self-legislation, remember? It may seem paradoxical, but an action can result in decreased autonomy (suicide being the extreme case) and still be perfectly autonomous. It need not be, of course, but it may be.

I would be remiss if I did not close on a more positive note. Dr. Macklin's paper is a sensitive treatment of a difficult problem. It cannot stand as her final word, but it makes one impatient to see her next effort.

PART III
Death and Dying
Electing Heroic Measures

Chapter 11

What is Heroic?

James A. Bryan, II

To be heroic is first to accomplish. It is to have valor. It is to be first. It is to defeat the enemy, to push back the frontier, to lead the charge. To establish the civilization, to surmount the odds, to lighten the darkness, to extend the faith, to keep the faith, to bring good, or to give life. It is being bigger than life. Sometimes just to live. But, it is to be remembered.

I was raised on heroes, but the line between just being and being remembered for what is accomplished sometimes gets blurred because as time goes on the fact of living at all seems heroic. If a life is accompanied by a memorable event, it is easier to identify that life. If a life reflects memorable character traits, or characteristics, ideas or accomplishments that generally reflect "good" or "better," then the approbrium "Hero" is more likely to be applied to that person.

When I think of heroes names come forward. Names that are bigger than life? By being bigger than life have they defeated death? If only a few remember—does that make for heroic? "Better a live coward than a dead hero!" "Don't kill yourself trying the impossible —don't bite off more than you can chew." or "Dream the impossible dream." All of us have heroes we would list together—the ancients from the Bible (Abraham through Paul including the gentle Ruth who is easier to remember than the woman Jael the Kenite who drove a tent peg through the skull of the sleeping Canaanite general), the ancients from our Western world, Alexander, Horatio, Caesar—was Brutus a hero—the gentle St. Francis, what about Augustine? Michelangelo, Galileo, Copernicus, DaVinci, Shakespeare, Bacon, Harvey, Erasmus, Boerhaave, Sydenham, Washington, Rush, Napoleon? Lincoln, Lee, Laennec, Virchow, Pasteur, Koch, Ehrlich, Livingston, Osler, Halstead, Kelly and Welch, Pepper, Roentgen, Curie, and Cushing. The litany could go on and on. If you noticed, I tailed off with a few medical names and stopped with the early part of this century. Maybe the identity of heroes should depend on a collective memory of over 100 years—I wonder how many of you remember the surgeon who did the first heart transplant or should you remember his patient—or should you remember the donor?

Can the identification of heroes define what is heroic in present time: Is there a difference between heroic and appropriate, or heroic and good? Today I propose to tell a few "war stories" from the current and past medical battlefields and hopefully stimulate some thinking about what we should remember—because what we remember as good, most of us want to do, and what we remember as bad most of us want to avoid.

What is good and what is bad? Life is good, death is bad (or is it?). Pain is bad and pleasure is good (or is it?) Function is good, disability is bad (or is it, if life is there?). Most of all love is good and hate is bad—but that is too deep for me.

Mr. G. is seventy years old, his wife died two years ago, and his Parkinsons disease proved very difficult to manage—so much so that his children finally arranged for him to go to an "intermediate care facility"—i.e., nursing home. I do not know how often they visited. In late August of this year he fell and broke his hip and was taken to his local hospital where it was pinned. A few days after the operations his intestines stopped functioning; nausea, vomiting, fever, abdominal pain, bloody diarrhea, and abdominal distention followed; and the X-ray picture was typical of ischemic or infarcted bowel. This is a usually mortal condition associated with diffuse vascular disease and *very* poor chance of survival if operation is undertaken. Death comes because the bowel wall dies, bacteria invade the bloodstream, and shock, vascular collapse, and death ensue. The family was so notified and gathered around the morning after the process began and was clarified. I was told that the patient "suffered" during that day. The family decided that "something had to be done" and transfer to the university hospital was arranged. (Is this where the heroes are in medicine—or is this just where heroic technology is present—what is the relationship between the two? I wonder if Sir Edmund Hillary got to the top of Everest because of modern technology—because of his "team" of climbers, because of himself, all of the above or none of the above—luck, fate).

When Mr. G. arrived in the Emergency Room he was in profound shock and within three minutes went into cardiopulmonary arrest. The ER team pulled the patient into the "crash" room and applied the "code 100" drill. An endotracheal tube was placed, artificial respiration was applied, intravenous lines were placed, medication given, X-rays taken, the shock was stabilized, surgeons were called, the situation was explained to the family (who were contacted for the first time since the patient entered the hospital). They still wanted "everything to be done" and so he went to surgery, and so he went to the surgical intensive care unit—where even yet (1 month later) attempts to "wean" him from the respirator have failed. He is being maintained with the respirator, with hyperalimentation and good nursing care, and has undergone a tracheostomy and drainage of an abdominal abcess.

I wonder what mountain there was to be climbed. I wonder if any of the people involved in that situation were thinking about being heroes, about heroics, or about ethical principles. I doubt it. I wonder if a hero ever thinks about being a hero. In the balance of good and evil, medicine in this particular situation relieved acute suffering in the patient to substitute chronic suffering in the patient and his family. Did the family need to suffer to assuage guilt for past abandonment of the patient. Why did the doctors (the system as exemplified in the emergency room drill) move through the life-saving maneuvers without asking why, to whom, to what end? The questions of why, to whom, and to what end are examined as the process continues, and are raised more and more as time passes and the doctors and their team members get to "know" the patient., The sometimes conflicting principles of "Do no harm," "Beneficence," and "Respect for human life" are very confusing when a specific patient such as Mr. G. is encountered.

To talk about principles in the same breath as heroics may be a contradiction of terms. Principles are rules or what is expected and hero's break the rules in order to accomplish something judged good. The military metaphor is the one that most quickly comes to mind (and perhaps it is very appropriate, given our mind set about "the last enemy"). If the odds are ignored and the risks taken, it is heroic.

One other principle that is needed in an ethically sound medical act is truly informed consent to the "heroics." Too often to obtain this in a meaningful way in many medical situations where "heroics" are indicated is to make heroics useless since the timing of the act is often critical. If the situation is not prolonged there often will be no situation at all.

It is very true as illustrated in the saga of Mr. G. that one act leads naturally to the next so that if an iatrogenic "harm" expresses itself (known as a "complication") it is almost impossible not to escalate the medical intervention to unusual and even extraordinary (if not heroic) levels in order not to have the original harm lead to death.

Definition of good outcome and the rules, guides, and principles under which this good takes place evolve out of group experience. In medicine the basic good or the fundamental goal that overrides, everything else is that of life or living. The problems come as the definitions of life or living are examined and as persons other than the patients themselves are asked to judge what therapeutic maneuvers represent the best choices for these patients.

McNeil and colleagues presented a paper in the December 21, 1978 *New England Journal of Medicine* in which they assessed the choice of treatment of patients with stage I or II bronchogenic carcinoma. An attempt was made to assay the patients attitudes toward risk-taking for prolonged survival. It happened that these patients had a mean age of 67 years and a median age of 69. The age range went from 40 to 82, and the study was retrospective in that the attitudes of the patients were examined in the abstract after the choice of therapy was made.

In bronchogenic carcinoma of this stage, therapy boils down to either surgery or radiotherapy. Thoracotomy with its anesthesia, particularly in persons having bronchogenic carcinoma, i.e., cigaret smokers with some lung impairment and possibility of heart disease, carries with it the aura of heroics. There is a real immediate danger of death, and though in the long run in the most favorable of surgical outcomes there is a slight advantage of survival, the survival with radiotherapy is better than surgery for the first three years. The surgical approach is even more costly as the patient ages.

The patients who underwent the treatment were given a series of hypothetical options to determine their willingness to undertake short-term risk for long-term gain. If the statistics derived indicated that they thought life during the next few years much more important than life many years later the patient were called "risk-adverse" as opposed to the opposite type of patients who were called "risk-seeking." Twelve of the fourteen patients so analyzed were risk-adverse, and even the "risk-seeking" patients expressed neutrality about risk after a few years. Remember that these were "older" patients. Are heroes not usually young? Is that why death in a child is so hard?

Every day doctors make choices in situations with uncertain outcomes. The question arises: "How good are physicians at knowing what their patients value most?" Are physicians generally willing to be more "risk-seeking" than their patients. On the other hand, do certain situations move physicians away from the "risk-taking," "heroic" mode for their patients and what leads to this. What ethical principles operate in either case.

In the medical model, heroics are relevant when true life ensues. Inappropriate heroics are wasteful at best and harmful to the patient at worst. The place where medical heroics are applied is in the hospital, and within the hospital in "intensive care" situations. In the past few years the harmful aspects of heroics have been high-lighted and the entire community is conversing about "patient autonomy," the "right to die," or "death with dignity." Recently Drs. Jackson and Younger of Case Western Reserve University presented some clinical situations from that University Hospital's intensive care unit that illustrated the possible conflicts between the move for patient autonomy and the right to die with dignity and the total (clinical, social, and ethical) basis for the "optimal" decision:

The first case was an 80-year old man with chronic obstructive pulmonary disease who entered the unit in respiratory failure requiring tracheal intubation and artificial respiration. He had survived a similar seige four years earlier that had required two months of respirator care. After two months of intensive care it became evident that continued artificial respiration was necessary. He would indicate to the doctors that he wanted the tube to be removed and "if I make it, I make it," whereas when the family was present he would indicate to them that he wanted maximum care even if it mean staying on the respirator forever. The staff naturally chose to honor the "life-sustaining" side of that choice and the respirator continued for two more months until a hospital acquired infection led to further problems and his heart fibrillated (no attempts to resuscitate were made).

A second patient, while being treated for a lymphoma, became acutely ill with metabolic and cardiac complications. While being treated for these he refused further treatment of his tumor; however, as these complications cleared, he became brighter in mood and asked that chemotherapy be resumed. His autonomy when exercised under depression led to an anti-heroic stance.

The third patient was a 52-year-old man who had retired two years previously because of physical disability from multiple sclerosis, though he remained active in his family's affairs. He was admitted to the intensive care unit in an alert state having attempted a drug overdose to commit suicide. He expressed his desire to die with dignity.—Is this not heroic? It appears though that the psychiatrist uncovered the fact that his withdrawal and depression coincided with his mother-in-law's illness which diverted his wife's attention away from him. A problem that he had "too much pride" to complain about.

Another patient was receiving chemotherapy for metastatic bronchogenic carcinoma when she had a sudden seizure and cardiac arrest. After resuscitation she was transferred to the intensive care unit in a comatose state, in shock and hypotensive with a flail chest and tension pneumothoraces. The mutual reaction of the medical staff was why torture this lady given her basic disease. Her course continued to be complicated and her survival was very questionable. Her daughter who was 7 months pregnant (1st grandchild of the patient) passed on to the staff the strong desire that her mother had to see the first grandchild. This supported increased efforts and the patient lived, was discharged from the hospital, and saw the child. Heroic!!

Let us now look at some clinical situations in which "nonaction" (nonheroic?) approaches have been put forth. Maybe these are heroic after all.

In 1977 Imbus and Zawachi wrote about the approach taken at the Los Angeles County–University of Southern California Medical Center Burn Unit toward patients with injuries so severe that survival was unprecedented. These patients were so judged based on clinical assessment of factors such as any combination of massive burns, severe smoke inhalation, or advanced age. This asessment would equal a certainly mortal designation of previous experience gathered nationally plus the experience of the local "team." "The most experienced physician is consulted, day or night to evaluate the patient." During 1975 and 1976, 748 dispositions were made from the center, 126 deaths occurred—18 children and 108 adults. Of the adults who died, 24 were diagnosed on admission as having injury without precedent of survival. "Twenty-one of these patients or their families chose nonheroic or ordinary (pain solution) medical care; three chose full treatment measures." No children were involved in this study. Graphic description of the differences in care and arguments for patient autonomy were made—either decision by the patient in this decision—i.e., "full effort" or "let me go" in my thought is heroic.

This year, Brown and Thompson wrote an article from the University of Washington entitled "Nontreatment of Fever in Extended-Care Facilities." In this study the medical records of 1256 people admitted to nine extended-care facilities in the Seattle area in 1973 were examined. Fever, defined as two temperatures of 101 to 101.9°F within 24 hours or one temperature greater than 102°F developed in 190 patients before two years of stay. "Active" treatment, defined as antibiotics or hospitalization (or both) was ordered for 109 of whom 10 died (9 percent). Active treatment was *not* ordered for 81 patients of whom 48 (59 percent) died. The pre-decision factors that showed a significant relation to such nontreatment were: diagnosis, mental states, mobility, pain and narcotics prescribed. Some other interesting associations to nontreatment included "relation of the physician to the patient." If it was the primary physician, oncologist or surgeon, 37 percent were untreated whereas if an alternate physician was contacted only 15 percent. Another variable proved to be the size of the facility where the patient was hospitalized. Twice as many patients in larger (50 bed) facilities received treatment as in the smaller facilities.

In reviewing this work it appears:

(a) That going to a hospital means treatment.

(b) That if life is done, i.e., there is deterioration, pain, and immobility, the doctor may be more supportive in "letting go." A heroic decision?

(c) The better the patient is known the more likely the nontreatment decision. In 18 patients the system changed from a treatment to a nontreatment approach.

Once again, this study was done on adults, those who have finished their race with the goal achieved. It is interesting that certain conditions (particularly cancer) were associated with the nontreatment decision. Variables that were not examined included measures of aloneness (i.e., the number and type of "nearest of kin") or economic or social status. The "heroic" measures applied by the medical committee to the late Francisco Franco come to mind. What a travesty!

Raymond Duff editoralizing in the journal *Pediatrics* recently wrote: "The problem can be stated as follows. Technology increasingly reigns while discourse about disease, and about patients, family and staff experiences and values is restrained. These conditions result from the public's and the profession's feelings that professionally designated, aggressive intervention to defeat disease and death is far superior to patient and family accomodations to disease and ultimately acceptance of defeat." He then refers to and champions two different types of heroism, "aspirational" and "humble," and argues that the best heroism is a balance in which both of these qualities operate.

Finally, I want to relate two situations from the distant past and recent past that illustrate the problems of achieving the best (heroic) result. In the mid-nineteenth century, a physician (Addison) described a clinical situation that included progressive wasting, weakness, darkening of the skin, inanition, then death. The autopsy showed the absence of the adrenal glands. They had been replaced by scar tissue. During the next fifty years, during one of the most exciting scientific periods, the adrenals were heavily studied. They were removed partially, totally, slowly, and rapidly. The fact that they were essential to life was established. The fact that they demonstrated at least two separate and distinct areas and functions was shown. Extracts were prepared from these areas and shown to be physiologically active in autologous species and heterologous species. Safe, i.e., nonfatal amounts of the extracts and routes of administration were established across species lines; yet, there had been no use in humans. One of the scientists reported these details then added that he had administered the extract of the adrenal medulla to a human with observations of the effects on blood pressure and pulse consistent with those observed in other animals. The extract of the adrenal medulla of course is adrenalin or epinephrine—the juice of heroes— the human on whom it was first administered was the scientist's teenaged son. I have always wondered, given this scientific story, what modern concepts of informed consent and experimentation on children would do to this scenario. The end of the story is a happy one with further step by step development of this most important principle. Maybe there are no heroes here.

The other situation is from the recent past and involves the modern treatment of childhood leukemia, which in the acute lymphatic variety promises a 50% cure rate if the proper "recipe" is followed. Recently a child was diagnosed as having the former condition, which was a universally fatal one and underwent initial treatment with chemotherapy. The protocol that was applied induced a complete remission as indicated by a repeat bone marrow examination and arrangements were made for consolidation of this remission by continued therapy, when the parents elected to withdraw the child from the protocol, as they perceived the treatment as worse than the disease. The physician in the case asked the courts to enjoin them against this decision, whereupon the family fled the state to take part in a therapy that has shown no scientific merit, and for which there is accumulating evidence of harm. The bad outcome in this case makes it most difficult for any heroics to be identified.

Heroic, superman, or just a human achievement of a goal against odds? Does science undergird all heroics in medicine and can a scientific approach to analyzing the outcome give a clue to us mere mortals what to do?

References

1. McNeil, B. J., Weichselbaum, R., and Parker, S. G. Fallacy of the Five-Year Survival in Lung Cancer. *New England J. Med.* **299**, 1397–1402, 1978.
2. Jackson, D. L., and Youngner, S. Patient Autonomy and "Death with Dignity": Some Clinical Caveats. *New England J. Med.* **301**, 404–409, 1979.
3. Imbus, S. H., and Zawacke, B. E. Autonomy for Burned Patients When Survival is Unprecedented. *New England J. Med.* **297**, 308–311, 1977.
4. Brown, N. K., Thompson, D. J. Nontreatment of Fever in Extended-Care Facilities. *New England J. Med.* **300**: 1246, 1979.
5. Duff, R. S. Guidelines for Deciding Care of Critically Ill or Dying Patients. *Pediatrics* **64**: 17, 1979.

Chapter 12

The Moral Justification for Withholding Heroic Procedures

Tom L. Beauchamp

In this paper I discuss moral justifications for withholding therapeutic procedures when it is known that persons will die as a result.
These justifications are commonly used both for cessation of the
procedures and for withholding all use of therapy. I construe all such
acts as instances of euthanasia.[1]

I shall approach these issues both constructively and critically.
Critically, I briefly outline four traditionally proposed justifications
for ceasing or withholding therapeutic procedures that are inadequate in some major respect (Section I). I then criticize the presently
operative policy statement of the House of Delegates of the American
Medical Association, which rests on a fifth, and widely accepted,
justification for cessation of treatment. I then offer some constructive
suggestions regarding the substantive criteria that determine when it
is not obligatory to prolong life.

I

Four common justifications for cessation of treatment about
which only brief questions can be raised here are the following:

1. *The Death Justification.* Therapeutic procedures may be withheld if and only if the patient is dead. This protective approach certainly states a sufficient condition for withholding therapy, but whether the condition is a *necessary* one is among the fundamental questions here under investigation.

2. *The Patient-Choice Justification.* Therapeutic procedures may be withheld if and only if the patient gives informed consent to withholding or rationally refuses therapy. This plausible proposal is not necessary and probably not always sufficient, as we shall see.

3. *The Physician-Responsibility Justification.* Therapeutic procedures may be withheld if and only if the knowledgeable physician judges in the light of medical evidence that they be withheld. This proposal may state a justified *procedure* for arriving at decisions concerning withholding, but the justifiability of a procedure is not a justification of a moral action.

4. *The Extraordinary-Means Justification.* Therapeutic procedures may be withheld if and only if extraordinary or heroic means of life sustenance are being employed beyond the bounds of ordinary treatment. It is not clear, however, what "extraordinary" and "heroic" means are; and it is unclear that the appeal to them functions as a *justification* for withholding, since it would remain to be shown wherein our obligation to the patient lie.

II

A fifth justification of withholding, and probably the one most commonly relied upon in medical tradition, is the following: withholding is permissible in a range of circumstances in which the patient is merely *allowed to die*, whereas it would never be permissible to *kill* the patient. The problem to be treated in this section is whether or not there is a valid distinction between active and passive euthanasia, a distinction that rests on the distinction between killing and letting die. I shall first deal with the problem as it emerges from a policy statement of the house of Delegates of the American Medical Association, adopted in December of 1973. That policy statement distinguishes between mercy killing and the withholding of treatment in order to allow a patient to die. Whereas mercy killing is said to be entirely contrary to "that for which the medical tradition stands," at least one form of withholding treatment is isolated as justified. It is said to be justified on the part of doctors to withhold or cease treatment in a narrowly restricted range of cases, viz., those where[2]:

1. The patients are on extraordinary means.
2. Irrefutable evidence of imminent death is available.
3. Patient and/or family consent is available.

When these three conditions are present, it is said that the physician is justified in allowing a patient to die. Our problem now is whether such advice is good advice for doctors, or confused and unjustifiable advice that stands in need of restatement.

This issue is as controversial legally as morally. In Judge Muir's initial verdict in the Quinlan case, he argued that "The fact that the victim is on the threshold of death or in terminal condition [or presumably is on extraordinary means] is no defense to a homicide charge."[3] This position works against the AMA position. Or more precisely, what the AMA has declared to be morally justified is not legally justified, if one reads the law as Judge Muir does. Resolution of this legal confusion seems to depend on whether turning off respirators and ceasing or withholding similarly important life-sustaining therapies are or are not human *acts* that lead to death. This problem too is one we shall be considering.

In a now classic article published in the *New England Journal of Medicine*, James Rachels advanced an argument against the policy statement of the AMA.[4] His case is based primarily on three arguments that may be summarized and analyzed separately.

1. Active euthanasia is in some cases more human than passive euthanasia.
2. The conventional AMA doctrine leads to euthanasia decisions on morally irrelevant grounds.
3. The distinction between active and passive euthanasia does not itself have any moral significance.

Rachel's arguments are intended to show that the AMA policy statement is conceptually confused and morally misleading. He finds it confused because the AMA stakes everything on the *bare difference* between killing and letting die. For the moment I shall develop this point in my own terms, which I freely admit involves some embellishment of Rachel's views.

Suppose, for illustrative purpose, that it is sometimes right to kill and sometimes right to allow to die. Suppose, for example, that one may kill in self-defense and may allow to die when a promise has been made to someone that he would be allowed to die. Here conditions of self-defense and promising justify actions. But suppose now that someone *A* promises in an exactly similar circumstance to kill someone *B* at *B*'s request, and also that someone *C* allows someone *D* to die in an act of self-defense. Surely *A* is obliged equally to kill or to let die if he or she promised; and surely *C* is permitted to let *D* die if it is a matter of defending *C*'s life. If this analysis is correct, then it follows that killing is sometimes right, sometimes wrong, depending on the circumstances, and the same is true of letting die. The *justifying reasons* for the action make the difference as to whether such an action is right, not merely the *kind* of action it is (killing, letting die, sacrificing, suiciding, or whatever).

This point can be clarified by considering a different distinction, and I choose for illustrative reasons the distinction between adultery and fornication., One might believe that fornication is permissible though adultery is not, but one might not believe that the *bare difference* between fornication and adultery itself accounts for this moral judgment. One might hold this view *because* one believes in the consequential view that adultery requires or leads to infidelity, while this consequence need not obtain in cases of fornication. Further moral reasons beyond the bare distinction between fornicating acts and adulterous acts, then, must be adduced in defense of one's moral position on the rightness or wrongness of fornication and adultery (value laden as all these terms may be).

Suppose, however, that fornication *did* lead to infidelity. In moral consistency one would have to hold that it would be wrong. And if adultery did *not* lead to infidelity, then in order to remain consistent it would have to be declared acceptable. Similarly, if letting die led to disastrous conclusions, but killing did not, then letting die, but not killing, would be wrong. Consider, for example, a possible world not too distant from one we may come to know. In this world dying would be indefinitely prolongable even if all heroic or extraordinary therapy were removed and the patient were allowed to die. Suppose that it costs over one million dollars to let each patient die, that nurses commonly commit suicide from caring for those being "allowed to die," that physicians are constantly being successfully sued for malpractice for allowing death by cruel and wrongful means, and that hospitals are uncontrollably overcrowded and their wards filled with communicable diseases that afflict only the dying. Now suppose further that killing in this possible world is quick, painless, and easily monitored. I submit that in this world we would believe that *killing is morally acceptable, but that allowing to die is morally unacceptable.* The point of this example is again that it is the circumstances that make the difference, not the bare difference between killing and letting die. And since the bare difference between killing and letting die—if there is a difference—is not a morally relevant one, we ought not to pass policy pronouncements on such a moral basis. The AMA statement (on this interpretation) is thus morally shaky.

III

I shall now sketch what I believe to be the most plausible justifications for withholding therapeutic procedures. My arguments are intended to cover withholding where heroic or exotic machinery is removed as well as decisions not to recommend or to discontinue more common medical treatments such as drugs. I do, however, wish to emphasize the limits of my account. First, it is only a proposal or suggested justification sketch, not a fully argued justification. Second, I do not maintain that my various conditions are both necessary and sufficient conditions of justified withholding. I maintain only that each set of conditions I propose is sufficient to justify withholding and that these conditions are relevant to practical medical situations in which it must be determined whether to withhold therapy. However, there may be other conditions that are also sufficient. Indeed there clearly is at least one other condition that is sufficient, viz., that the patient is dead (justification #1 above). What I do, then, is to offer justification sketches, each of which offers a set of sufficient conditions. One set is for refusal of treatment constituting voluntary euthanasia, while two are for refusals of treatment constituting different types of involuntary euthanasia.

Voluntary Euthanasia. According to the "Patient's Bill of Rights" of the American Hospital Association, "The patient has the right to refuse treatment to the extent permitted by law and to be informed of the medical consequences of his action."[5] There is now ample evidence that the opportunity to refuse therapy is of great importance to many competent patients, who believe deprivation of choice constitutes a serious loss of liberty that is psychologically damaging.[6] In such cases I would hold that a patient request, when informed and rational,[7] is sufficient to justify ceasing therapy. Any other general theory will lead to unnecessary cruelty and deprivation of liberty. The physician is responsible for what he or she does, but is not responsible for making the decision for the patient. The physician can remove himself or herself from the case, but the only adequate grounds he or she could have for not complying with patient wishes (or for not removing himself as the attending physician) would be moral or legal grounds that are in *all* similar cases of sufficient weight to limit liberty.

An example of a case where seriousness might reach this extent emerged in the infamous *Georgetown College Case*. Here one reason cited by the court for forcing a transfusion on an unwilling patient was the following: "The patient, 25 years old, was the mother of a seven-month-old child.... The patient had a responsibility to the community to care for her infant. Thus, the people had an interest in preserving the life of this mother."[8] It has also been argued that the child, even more than the community, is seriously damaged by the mother's decision.[9] Although I am unsure of the actual status of obligations in this dilemmatic case,[10] it constitutes the sort of case where pressing needs might be sufficient to override patient choice. Though I shall not argue the point here, I believe that the person's individual liberty can be justifiably overridden only for a moral reason that pertains to the welfare of others and not for paternalistic reasons.[11]

Involuntary Euthanasia. I turn now to the justification of involuntary euthanasia where death results from cessation of therapy. The interesting cases here almost always involve either (a) conscious persons who are so debilitated that they are incapable of making an informed choice, and probably even of being minimally informed, or (b) persons in a "persistent vegetative state," to use the language of the Quinlan trial, and who are (predicted to be) irreversibly comatose. Although (a) and (b) involve significantly different cases, withholding or cessation can be justified in both when certain conditions are present; and the justifying conditions are rather similar in both instances.

I shall consider the less difficult class of (b) cases first. Here the AMA position is promising, though misleadingly formulated. As we have seen, this position requires that "there is irrefutable evidence that biological death is imminent." The AMA policy statement assumes, and I share the assumption, that once a case is entirely hopeless there is no compelling moral reason for continuing treatment because *life under such conditions* is not of sufficient value to compel its sustenance. I do wish to emphasize, however, that this moral premise is being assumed and not argued for. Adequate resolution of the issue of whether life is more valuable under such conditions than this assumption admits would involve consideration of theories of personhood and of the principle of the sanctity of life; and such considerations would here take us too far afield.

On the other hand, the AMA statement just quoted is objectionable for both epistemological and moral reasons. Because empirical evidence is not in principle irrefutable, no more could be demanded than that it be sufficient evidence, as judged by prevailing medical criteria. Second, "imminent" is an all too indefinite term. One might say that another war in the Mideast is imminent, meaning "in less than two years." Or one might say that a person's death is imminent, meaning "any second now." The AMA perhaps landed upon such an ambiguous word because it allows physicians latitude of judgment. I am not against latitude here, but in their eagerness to advance a moral view the AMA simply missed the target at which they were aiming—or should have aimed. The important factor is the irreversible character of the patient's condition, not the imminence of eventual certainty of death. Karen Quinlan's death was not imminent in at least one interpretation of that term, and imminence played no role in either Judge's opinion. Imminence simply is not relevant in the way irreversibility of a terminal disease or condition is relevant. If death were *not* imminent, but the comatose condition was irreversible, imminence would play no role in our judgment concerning what ought to be done.

A condition's being irreversible and its being *terminal* are also different. One can have an irreversible condition that does not involve a terminal disease (at least on widely accepted theories of the nature of disease). I suggest, then, that we distinguish between the conditions of terminal illness and irreversible loss of mentation. Now for the moment we are considering only persons in a persistent vegetative state. Here it is precisely the condition of irreversible loss of mentation that is relevant, not terminal illness. However, even if irreversibility is a necessary condition of justified cessation, one qualification on the use of the term "irreversible" should be added. "Irreversibility" in the present context means *predicted* irreversibility. But not just any prediction will do; it must be prediction in the light of the best scientific information available to the person making the prediction. This information would include data about research protocols and possible "emerging" therapies that might become available. A case ceases to carry the label "irreversible" if a promising therapy looms on the horizon.

The analysis thus far suggests that two conditions must be satisfied for justified withholding or cessation of therapeutic procedures in the (b), or "persistent vegetative state," range of cases:

> (1) Irreversible loss of mentation (justifiably predicted).
> (2) Unavailability of new therapies (justifiably predicted).

When these two conditions are satisfied, we are usually justified in ceasing life-prolonging therapeutic measures, whether heroic or ordinary. Whether we would also be justified in using some active means other than (passive) withdrawal is a further question, but at least we are usually justified in cessation of or withholding of procedures.

We are only *usually* justified in such cases because these two conditions are not always jointly sufficient for cessation. The reason derives from our earlier discussion of patient choice. I previously insisted on the importance of respecting a patient's autonomy, even if the act constitutes euthanasia. In order to be morally consistent, we would have to give serious consideration to a patient's autonomous wish in the present context. Suppose a perfectly competent person regards death as the most serious evil that can befall a person and leaves explicit directions that he or she does not want to be "terminated" even if conditions (1) and (2) obtain. I would argue that the patient's decision here, as before, ought to be honored. It follows that a third condition must also be present for justified cessation:

> (3) No applicable request has been made by the patient, when informed and competent, that therapy should be continued if conditions (1) and (2) are satisfied.

Although there is a prima facie obligation to comply with any such patient request, I would not argue that there is an absolute obligation. Many other considerations might be overriding.

I turn now to the justification of (a) cases that involve conscious persons so debilitated that they are irreversibly beyond the bounds of making an informed choice, and probably even of being informed. For such incompetents, there is no irreversible loss of mentation. Rather, there is a condition of impairment of function that renders the person incapable of informed judgment. In these cases, at least the following three conditions, which parallel the aforementioned three, are *necessary* conditions of justified cessation of therapy (assuming "justifiably predicted"):

(1) Irreversible loss of the capacity for informed judgment.
(2) Unavailability of even a promising curative therapy.
(3) Nonexistence of patient request that therapy be continued.

Despite these conditions being necessary for justified cessation, they are not jointly sufficient [as were the three parallel conditions for (b) cases]. There are countless examples to show that they are not sufficient. For example, there exist cases of nonepisodic, senile brain disease of irreversible etiology where there has been brain shrinkage, scarred cells, and so on. This condition in turn produces senile psychosis. Still, the person so afflicted may be able to enjoy meals, find pleasure in a back rub, and so on.

The critical question is whether there is a fourth condition that, when added to the first three conditions, creates a set of jointly sufficient conditions of justified cessation? I think there is, and I suggest the following as its formulation:

(4) The patient requires a life-sustaining therapy that would itself produce, on balance, greater suffering (or other harm) for the patient than would be produced by not using the therapy.

There are few more important decisions than those to end a therapeutic procedure, which in turn ends a life. But in cases of the sort we are considering it would be unjustified to continue therapy knowing that it is tantamount to the production of a greater balance of suffering for someone who is incapable of choosing for or against such therapy. An interesting example, though not one involving cessation of therapy, occurred in the case of *In re Nemser*, where a judge refused to require the amputation of the foot and ankle (a transmalleolar amputation) of an eighty-year-old woman whose competency was dubious and where her children sought to give a proxy consent to the operation.[12] It was medically unclear how much benefit would occur, though substantial suffering was certain. The operation might actually have cured her, but the substantial hazards and suffering involved were judged sufficient to override the possible beneficial outcome (though Judge Markowitz refers to the decision as "an example of a grave dilemma").

Two more recent cases are perhaps more applicable, however. In a critical 1972 case on informed consent requirements, *Canterbury v. Spence*, it is said that a physician's obligation to proceed with therapy in emergency cases where persons are incapable of consenting depends upon whether the "harm threatened by the proposed treatment" is or is not outweighed by "a failure to treat."[13] More recently the Supreme Judicial Court of Massachusetts held in *Superintendent of Belchertown v. Saikewicz* that the best interests of an incompetent but conscious person are not always served by forcing therapies on them that reasonable competent persons might well refuse. The court and a guardian refused in this case to elect a treatment for leukemia that might have brought about temporary remission. Their grounds were that the illness was ultimately incurable, that chemotherapy added significant discomfort and adverse side effects, and that Mr. Saikewicz could not understand what was being done to him.[14]

I conclude from this analysis that conditions (1–4) form a set of jointly sufficient conditions of justified cessation of therapy in cases of the (a) sort.[15]

Finally, there is yet another class of persons different from those we have thus far considered. These are persons who are capable of informed judgment, but who have serious illnesses and are enormous social and medical problems. I have in mind persons who need such expensive equipment that society cannot in a realistic sense afford it, persons who irresponsibly use (or fail to use) valuable resources such as kidney dialysis machines, and persons who create havoc in a hospital by destroying equipment, manhandling nurses, and so on.[16] Here I shall only make two brief comments. First, to treat this problem comprehensively issues of distributive justice involved in the allocation of scarce medical and other social resources would have to be pursued, and this enterprise would probably also require discussion of the justifiability of triage and deselection criteria. This problem has never been adequately handled and obviously cannot be undertaken here.

Second, I know of no plausible set of conditions except those involving limited resources that would justify withholding or cessation of therapy without patient consent under these conditions. Here we may be tempted by act-utilitarianism into saying that if the greatest good for all concerned resides in withholding or stopping therapy, then it should be withheld. But innumerable counterexamples confront such a suggestion. They show clear violations of basic human rights, and it is precisely this sort of conclusion that gives utilitarian-ism the (I think unjustified) reputation of being a monstrous moral theory that runs roughshod over human rights. I am myself sceptical that it would ever be justified to cease therapy for such persons except in cases where there is a strong conflicting obligation that requires cessation or withholding of therapy.

I conclude that patient choice generally justifies cessation of therapy and that there are three conditions of justified cessation for the irreversibly comatose (b), and four for the irreversibly debilitated but conscious incompetent (a). If there are other conditions for other classes of patients besides those who are dead, I am unaware of them.[17]

Notes and References

1. "Euthanasia is a word that can be used in a variety of ways. I believe, however, that a descriptive, non-value-laden defi-nition can be provided. Arnold Davidson and I have argued for the following definition of euthanasia in "The Definition of Eutha-nasia," *The Journal of Medicine and Philosophy* **4** (1979), pp. 294–312:
The death of a human being *A* is an instance of euthanasia if and only if: (1) A's death is intended by at least one other human being, B, where B is either the cause of death or a causally relevant feature of the event resulting in death (whether by action or by omission); (2) there is either sufficient current evidence for B to believe that A is acutely suffering or irreversibly coma-tose, or there is sufficient current evidence related to A's pre-sent condition such that one or more known causal laws supports B's belief that A will be in a condition of acute suffering or irreversible comatoseness; (3) (*a*) B's primary reason for intend-ing A's death is cessation of A's (actual or predicted future) suf-fering or irreversible comatoseness, where B does not intend A's death for a different primary reason, though there may be other relevant reasons, and (*b*) there is sufficient current evidence for either A or B that the causal means to A's death will not produce any more suffering than would be produced for A if B were not to intervene; (4) the causal means to the event of A's death are chosen by A or B to be as painless as possible, unless either A or B has an overriding reason for a more painful causal means, where the reason for choosing the latter causal means does not conflict with the evidence in *3b*; (5) A is a nonfetal organism.

2. The full AMA statement appeared as follows: "The intentional termination of the life of one human being by another—mercy killing—is contrary to that for which the medical profession stands and is contrary to the policy of the American Medical Association. The cessation of the employment of extraordinary means to prolong the life of the body when there is irrefutable evidence that biological death is imminent is the decision of the patient and/or his immediate family. The advice and judgment of the physician should be freely available to the patient and/or his immediate family."

3. Judge Muir, *In the Matter of Karen Quinlan*, Vol. I. Arlington, VA: University Publications of America, 1976.

4. Rachels, James. "Active and Passive Euthanasia," *New England Journal of Medicine* **292** (January 9, 1975): pp. 78-80.

5. Reprinted in Beauchamp, Tom L., and Walters, LeRoy. *Contemporary Issues in Bioethics*. Encino, Calif.: Dickenson, 1977, Chapter 4.

6. Cf. the cases in "Informed Consent and the Dying Patient," *Yale Law Journal* (1974), pp. 1658 (fn. 139), 1660.

7. Whether consent is informed and rational is, of course, often difficult of ascertainment. This problem, including mention of conditions under which a patient request ought to be refused, are interestingly discussed in Kenny F. Hegland, "Unauthorized Rendition of Lifesaving Medical Treatment," in *California Law Review* 53 (August 1965). Cf. esp. pp. 864f, 877.

8. Byrn, *op cit.*, p. 33. Also, the issues are nicely presented in *In re Osborne*. On some interpretations of the decision in *Georgetown College* the verdict was reached as much on grounds of questionable competence as on the utilitarian grounds I have cited.

9. Cantor, Norman. "A Patient's Decision to Decline Life-Saving Medical Treatment: Bodily Integrity Versus the Preservation of Life," *Rutgers Law Review* **26** (1973): pp. 228, 251–254.

10. Hegland takes the view that society has sufficient interest in the life of the patient in this case to "deny him a legal right to refuse lifesaving treatment," *op. cit.*, p. 872.

11. A case that I regard as paternalistic and unjustifiable for this reason is *John F. Kennedy Hospital v. Heston* 58 N.J. 576, 279 A.2d 670 (1971). A clear reversal of this way of thinking is the appellate court decision in *In re Estate of Brooks* 32 Ill. 2d 361, 205 N. E. 2d 435 (1965), though the latter preceded the former by six years. In the former case, see esp. pp. 584f, and in the latter esp. pp. 440ff.

12. In Katz, Jay, ed. *Experimentation with Human Beings*. New York: Russell Sage Foundation, 1972, p. 709.

13. *Canterbury v. Spence* 464 F 2d 772 (1972). Though the formulation in this case is close to my third condition, I would lay emphasis on suffering rather than on so flexible a word as "harm."

14. *Superintendent of Belchertown v. Saikewicz*, Mass. 370 N.E. 2d 417 (1977).
15. I might add, as an addendum to this analysis, that it could be questioned whether *cessation* is a justifiable means in some cases, since the argument I previously adduced in Section II, when coupled with the argument here in Section III, would lead to the conclusion that in some cases an *active* means is preferable to a passive means. If so, a fourth condition would be needed specifying that active means should be used if a passive means would produce greater suffering or harm for the patient than would be produced by cessation of the therapy. This proposal, however, cannot be easily formulated without careful qualifications, since it could easily lead to a requirement of murder for an insufficient reason.
16. A case of thoroughly irresponsible use is found in "Long-Term Dialysis Programs: New Selection Criteria, New Problems," *The Hastings Center Report* 6 (June 1976), p. 9.
17. This conclusion does not entail the existence of a "right to die." This peculiar "right" seems to me to carry dangerous breadth of application, and I would be opposed to my arguments being used to support arguments for such a right.

Chapter 13

Response to Bryan and Beauchamp

Kevin Lewis

I must confess that my interest in health care, biocthics, and the agonizing dilemma of one's personal decision to request or refuse extraordinary means of life support is primarily that of your average citizen.

I do believe that there are matters of all kinds that bear scrutiny from an interdisciplinary perspective. The matter of "hcroic" intervention to save and prolong human life is certainly one of them. And I have decided that I do have something to say, as a cultural historian with particular interest in religious experience and theological reflection. (Here I apologize for throwing you into yet another realm of professional jargon).

I owe to James Bryan the inspiration for these remarks, and I offer them in the way of an excursion beyond the provocative question with which he left us, "What is heroic?" I decided to think this through for myself and in my own terms. (This question was answered, after a fashion, just 200 years ago by Samuel Johnson, who observed to Joshua Reynolds that "He who aspires to be a hero must drink brandy.")

It struck me that I have never questioned the origin of the use, by the medical profession, of the term "heroic" to refer to life-saving,

163

life-preserving (and life-prolonging) measures involving sophisticated technology and high cost. I am grateful to Bryan for suggesting, by implication, that we might at least try to define the term in order to use it with some consistency, and better yet, in such a way that reflects our understanding of its historical meaning. Why *did* the term "heroic" seem appropriate in the first place? Some of you may know the history—I don't.

I have done a small amount of re-reading, and I will warn you that two of my texts for this Response are: Thomas Carlyle's *On Heroes and Hero Worship* (1841) and Kierkegaard's *Fear and Trembling* (1844), both from the century of John Stuart Mill.

One finds, with a little research, that, for the greater part of Western history, the notion of the hero entailed the notion that certain men (but probably not any women) stand out from the crowd conspicuously by virtue of the favor shown them by the gods, by virtue of miraculous powers or consciousness invested in them by divinity. Guenter Risse has called our attention to the functions of the Healer-Shaman-Priest of primitive societies. When one traces the idea of "heroism" in history, one discovers that, however more sophisticated the job description becomes, the "hero" retains the attributes of the Magician-Priest generally speaking. He is a mortal who partakes of immortality. After death, the "hero" comes to be venerated, if not actually worshipped, as half-human, half-god.

The status of religion having come to be what it is in our society, it is no wonder that we look in vain today for heroes comparable to the heroes of ages past. This is one logical consequence of secularization.

Here is Carlyle's Romantic definition:

> The Hero is he who lives in the inward sphere of things, in the True, Divine, and Eternal, which exists always, unseen to most, under the Temporary, Trivial: his being is in that; he declares that abroad, by act or speech as it may be, in declaring himself abroad. His life, as we said before, is a piece of the everlasting heart of Nature herself: all men's life is,—but the weak many know not the fact, and are untrue to it, in most times; the strong few are strong, heroic, perennial, because it cannot be hidden from them.

This is both "Romantic" *and* "religious."

I propose that when we label extraordinary life-saving-prolonging procedures "heroic," we are doing so *without* consciously recalling the etymology of the word, but *with* what I sense to be an ambivalence that reflects our awareness that, by using these artificial measures, we are indeed "playing God" (invading God's space) and, by doing so, putting ourselves at risk spiritually.

Now, having once affixed the label "heroic" to this scale of medical intervention, it does make equal sense, as Bryan is suggesting, to label the refusal of these measures "heroic" also. For in either case, no matter which way one is now able to choose, one is entering a sphere of choice and behavior that traditionally has been understood to hold "religious" significance, a sphere whose precincts traditionally have been guarded by the priestly adept who manipulates the mysteries.

Today we quite naturally understand these decisions to be essentially ethical, moral decisions. And, of course, the impossibility of establishing acceptable guidelines for these agonizing, personal decisions is our great dilemma. It is appropriate for ethicists and bioethicists to search for a principle or a norm or norms that, when found, might ease our dilemma. (This is what ethicists *do*.) What we need, obviously, is a universal law that would command the respect of the total community and put an end to the conflicts into which we are thrown by our cultured moral relativism.

Here I propose that there is no such universal law or principle, and never will be. And that we will only increase our frustration by hoping that such a law will eventually be established.

Furthermore, keeping in mind now the historical meaning of the term "heroic," I propose that these moral dilemmas do indeed have a religious dimension, and that we have virtually, arguably admitted as much all along by our continued use of the term "heroic." I suggest that the religious dimension in this whole matter is unavoidably significant, and that, by trying to avoid it, we increase frustration and complicate the dilemma of decision.

This is not meant to be a hectoring, bullying appeal. I mean that we are discussing life, death, and decisions about matters that, throughout the past, were considered to lie in the hands of the gods. Whether a transcendent reality of a kind any of the major religions claims to exist *actually* exists is not at issue. What is important is the worldly human power of religious belief.

Here is my application. Ultimately, there is a crucial difference between God's law and human law. (I am speaking symbolically.) The behaviors and decisions enjoined upon individuals and communities by each of these two "laws" or law codes, respectively, will at some point conflict. This is a point of principle.

Kierkegaard became obsessed by this conflict and its effect on human lives, including his own. The Lord tells Abraham to sacrifice

Isaac his only son, his greatest treasure, to prove his faith. Abraham demonstrates to the Lord that he is prepared to do so. Had he done so, had the Lord not stopped him in time, Abraham would have become a murderer deserving of the appropriate punishment under law. He would have become a murderer, *but*—and Kierkegaard wrote a whole dissertation on this "but"—but, in obedience to God, he became a perfect "Knight of Faith," a religious exemplar-hero. The price of course is terrible. The world will never understand. For Abraham has gone beyond the moral law. He has shown that he is prepared to break the moral law if and when God so commands. And this is what it may mean to be religious, Kierkegaard reminds us: to pose a threat to the community and to risk punishment under law insofar as one obeys his or her conscience.

My conclusion is that one can only hope to make an "heroic" decision of a kind that will give one's conscience peace by consulting what's left of one's *religious* conscience. The moral law, in these matters, can give no peace. There is no one "right" decision, not for us. There is only what one chooses to do, by one's own lights, consoled by one's religious beliefs, if one has them, or simply not, if one doesn't.

And we have been dimly aware of this situation as I have characterized it in our continued use of the term "heroic."

If you object, as well you might, that "heroism" means something *now* quite different from what it meant for all those centuries during which religious belief rested secure, I will agree, and so will Kierkegaard. For he insists that the "Knight of Faith" cannot *be* a hero. For a hero must be heroic in the eyes of others. Heroism arises out of the endorsement given the hero by his society. Since no one but God can understand or value Abraham's actions, since Abraham must be prosecuted as a criminal, as a murderer without justification, he cannot be termed a "hero."

"Heroism," so understood, becomes a quality confined strictly to the plane of acceptable moral behavior, of conformity to the erected moral standard, to the categorical imperative. But on this plane, in this matter, despite the clarity Tom Beauchamp has brought us, I submit, we are lost. So perhaps "heroic" is the wrong word. If so, lets find another.

PART IV
Advancing Reproductive Technology

Chapter 14

Ethics and Reproductive Biology

Daniel Callahan

The topic of "Ethics and Reproductive Biology" is exceedingly broad, complicated by the fact that, in our society, we possess no simple and comparatively common language for talking about a whole cluster of issues concerning reproduction. My approach will be to accept the broadness of the topic, and then try to analyze different components.

Until very recently in human history, it was taken for granted that there was comparatively little human beings could do to control procreation. They could neither control the number of children they had, nor control the genetic quality of their children. The process of procreation itself was only vaguely understood, and it was not until the 19th century that there was any clear understanding of the biological process of procreation and reproduction. Although it was vaguely understood that intermarriage between those closely related produced offspring with genetic defects, that was about all that was understood.

As a result, I believe that most human beings over most of human history understood procreation to be an essentially unalterable part of human life. Practically all of the great world religions, and the general mores of most societies, looked upon procreation as something akin to a sacred process, not to be tampered with. Moreover, the whole domain of sex and procreation was bound about by moral rules and customs, most of them negative in their orientation. Also there have been comparatively few societies that did not have moral and legal rules against abortion and infanticide, and the fact that one can here and there find exceptions only underscores the pervasiveness of the general moral and, until recently, legal rules.

It is hardly surprising that this broad area of human life was surrounded by very strong taboos and a great number of sanctions. Agricultural life placed a very heavy premium on raising enough children to assure the ability simply to till the soil and to keep alive. Moreover, a very high infant mortality rate meant that it was necessary (and in many places around the world today, it is still necessary) to produce a large number of children in order to guarantee that a few would still be living by adulthood. The fact, too, that there were very few alternatives for women except those of child-bearing and child-rearing merely added to the social pressures to produce children, and to the imposition of moral standards that would not stand in the way of that necessity. At the same time, the fact that primitive contraceptive techniques were usually ineffective, and sometimes dangerous, meant that there was little hope that either couples or societies could do much about controlling the number of children, even if they wanted to.

In short, the social and economic pressures throughout most of human history were all in the direction of encouraging, even requiring unlimited procreation. In that context, infertility became a stigma and when there was no significant role for women other than child-bearing and child-rearing, to be infertile was a very important social deprivation and normally led to the creation of an important social stigma as well.

Now if all of that was true concerning the number of children, an analogous situation could be found concerning the genetic quality of children. Practically nothing was understood until very recently about the causes or conditions of genetic or other defects. Hence, defective children were simply born, and there was nothing parents could do to control the matter at all. Moreover, because the whole process was mysterious and seemingly subject to few observable natural rules, a great deal of fear, anxiety, and superstition attended the birth of defective children and, more generally, characterized societal attitudes toward defectives. Such children for the most part were highly stigmatized and the parents who had borne them often felt themselves stigmatized as well.

Now human history seems to make evident that when people have little control over natural events, they also have a kind of paradoxical tendency to sanctify or endow that situation with the status of something not only beyond their control, but something perhaps that ought to be beyond their control. One might call this "fatalism," but the general assumption is that one ought to be fatalistic because the way things are is probably the way things ought to be.

What makes our present situation so traumatic is that, first, our new technological powers in the realm of reproduction enable us to do things that were simply thought utterly impossible by earlier generations. And, second, that the new powers have simultaneously forced us to reconsider in a most fundamental way the whole question of our relationship to nature, particularly our relationship to our own human nature. Third, that as an immediate consequence of that reconsideration, there has been an enforced recognition that it is by no means clear any longer just what is the wisdom of nature, if such there is, and much less clear what the Will of God might be concerning such issues. Thousands of years of human thought and experience seem all of a sudden to have been turned topsy-turvy, and perhaps of all the contemporary bioethical issues, the range of ethical problems associated with reproduction can be said to be genuinely new and unprecedented.

Let me briefly describe, in almost catalog fashion, some of the present technological realities, and some of the future possibilities bearing on procreation. Some of the present realities are these: that there are reasonably effective contraceptives, but not yet perfectly safe contraceptives, much less perfectly safe, cheap, and guaranteed contraceptives. Nonetheless, the prospect is that these will eventually be developed. Abortions are now safe, at least in early pregnacy. There is now a fairly decent knowledge of genetics; and by making use of family histories in particular, it is quite possible under a large number of circumstances to tell couples prior to marriage and procreation what the odds are that they will bear a defective child; and of course that knowledge is all the stronger if they have previously borne a defective child. This knowledge is of course by no means complete, however, and for the most part little is still understood about the etiology of many conditions leading to the birth of defective children. Through the use of prenatal diagnosis, it is now possible to detect an increasing number of genetic defects *in utero*— Down's syndrome and Tay-Sachs disease would be obvious examples. And the list of detectable genetic conditions *in utero* is expanding all the time. When those conditions are diagnosed, it then becomes possible for a woman to choose an abortion and try again for a normal child. Finally, I would mention as a present reality mass genetic screening for some conditions.

If those are some of the things we already have, there are a number of future possibilities on the way. One of these would be a very simple method of sex selection, prior to intercourse. Prenatal diagnosis can already be used for sex ascertainment *in utero*, but sex selection prior to procreation would simply be the next step, and it is most likely that relatively simple methods will rapidly become available. There are also increasingly simplified methods of prenatal diagnosis on the horizon, including already the fetoscope, which allows a visual inspection of the fetus.

It is likely that the whole process of genetic screening will develop rapidly and no doubt move to the use of simple automated tests, requiring only a blood sample. The use of in vitro fertilization —to which I will return—has been carried out in at least one and perhaps two or three such cases, and seems likely to continue to develop as well. As we look even farther down the road, the possibility of an artificial placenta is by no means impossible, and there are at least a few researchers working to develop it. There is also a strong possibility—through it may be decades off—of the development of techniques of genetic surgery, techniques that would allow for the correction of genetic defects while the child was still in the fetal state. Of course there has also been a considerable discussion of that most exotic of issues, cloning, but if that is to come, it is not likely for some time.

When one talks about "reproductive biology," one is thus really talking about a whole range of possibilities for the future. Since the past seems to be a relatively reliable guide in the whole field of science and technology, one can assume there will be constant developments, both theoretical and applied, in the field of reproductive biology.

Though it is possible to talk very broadly about the ethical problems of reproductive biology, it is probably a mistake to believe that all of the present realities and the future possibilities raise exactly the same kinds of ethical issues. Perhaps life, particularly the moral life, would be simpler if they did. But when one begins moving through the various present realities and future possibilities, it soon comes clear that many distinctions must be made, and that we are faced here with the multiplicity of issues, not one only.

Perhaps the broadest question that must be faced, and perhaps the question that does cut across all of the various realities and possibilities, is that of the human right to intervene in nature, or—to avoid the language of rights—the wisdom of radical interventions into actual processes. Of course, in many ways, that issue has long since been resolved. We approve of medicine's efforts to intervene and control those natural conditions that cause illness and disease, and we long ago realized that it is possible—as in the case of eye-glasses—to enable people to better adapt to their environment, and to adapt to some genetic shortcomings.

In short, the principle of intervention into nature to improve the human conditions is now well-established, and perhaps at that broadest level the question for us is: how far do we want that intervention to go, and at what point does it cease to be wise and prudent?

Beyond that broad question, I think it is possible to helpfully break down some of the technical developments into three general categories:

The first category would be simply the control of the number and spacing of children, and here of course one sees all the issues of contraception, abortion, sterilization and, more broadly, issues of population limitation.

The second large category of issues turns on the control of the genetic quality of children, and here one raises and runs into moral problems of genetic counseling and screening, prenatal diagnosis, sex selection, and, ultimately, genetic engineering.

The third category I would call the control of the process of procreation and gestation, and here one sees moral problems of AID and in vitro fertilization, and perhaps in the future surrogate mothers and artificial placentas.

These very general categories raise different types of ethical issues, though there are some moral problems that may cut across all of them. However, if we agree that intervention into nature is appropriate—as mentioned above—in some very general sense, then that forces us to move on to another class of issues, and to ask a question that should perhaps be asked of all of the three categories mentioned above: what will be the general consequences of such interventions? One can ask about the consequences for society as a whole and, particularly in the genetic area, questions about the genetic consequences for the human race. One can also ask about the consequences for families, for couples, and for individuals. One can also ask, even if one has accepted very generally the principle of intervention, whether there ought to be some moral limits to that intervention. Are there some moral principles available, or that should be developed, that ought not to be overriden even if one could show generally beneficial consequences? For instance, are there some basic human rights or principles concerning individual integrity and dignity that ought to bear on our judgments in this area?

Obviously, also, there is a broad cluster of questions concerning the process of decision-making: who ought to make such decisions concerning these interventions: individuals, families, or society as a whole?

It seems obvious that when one raises such a broad cluster of questions of that sort, one may arrive at very different types of answers for each of the three categories noted above. There is no reason, I would reiterate, to assume that there can be general answers. The problems with contraception, abortion, and steriliza-tion appear to raise very different issues than questions of prenatal diagnosis.

Perhaps if would be useful to look at in vitro fertilization as a case study of the ethical problems of reproductive biology. Compara-tively speaking, there has been less discussion of the moral problems of intervening into the process of procreation and gestation than has been the case with the question of the control of either the quantity or the quality of children. Perhaps there has been less discussion of a serious sort on those issues because the media tends to have focused on other issues, but perhaps also because the technology here has lagged the most. It was widely assumed, over a decade ago, that in vitro fertilization would come quite rapidly. That was not the case, and the work in that area went on for many years before the break-through was made. Moreover, in vitro fertilization is the first truly significant technological breakthrough in the area of procreation and gestation.

It is important to look briefly at the history of the debate over the technique. In the late 60s and early 70s a number of researchers announced that they were attempting to carry out in vitro fertiliza-tion. These included Drs. Edwards and Steptoe in England, and Landrum Shettles in the United States. The main reason they gave for carrying out this research was that they thought it would be a very effective means of relieving the infertility of at least some women. Now, interestingly, in the late 60s and early 70s, the initial response to this line of research was a general outcry against it, an outcry within the biomedical community and outside of that community as well.

Some of the response to the research seemed to represent a kind of visceral reaction, as if in tinkering with something as fundamental as procreation, some kind of important boundary line was being crossed. Hence, many of the objections at the time, while strong, seemed inarticulate, vague, and apparently more the expression of anxiety than a well-thought-out rational response.

However, one major line of objection was that the research was likely to be harmful, if not to the parent, certainly to the fetus so conceived. Many pointed out that this was a form of human experimentation, that is to say, experimentation with a fetus, to see whether a fetus could be developed by means of in vitro fertilization. Many argued that there was no way morally even to gain that knowledge. The only way one could find out whether the procedure was safe, was to carry it out in the situation where it was not known in advance and where hazards would be imposed upon the fetus.

One important result of the outcry and resistance at that time was that the research more or less went underground. The researchers who had spoken so enthusiastically and publicly about the issues in the late 60s and early 70s simply stopped talking. After about 1972, very little appeared in the scientific literature on the subject. Nonetheless, we now know that the research was continuing. Though the work began at least back in the late 60s, it turned out to be a far more difficult technical process than the researchers or others envisaged. The main technical problem was not fertilization of the egg outside the womb, but that of implanting the fertilized egg in the womb; that apparently proved to be very difficult, and it would appear that it is still by no means simple.

Nonetheless, in the summer of 1978, Louise Brown was born in England, and Drs. Edwards and Steptoe, who had helped to bring this about, basked in a great deal of limelight, as did the parents and the baby.

One might ask, what then, can now be said about their particular success? For one thing, the birth of the baby would seem to provide some evidence that the procedure was not as harmful as many suspected it might be. Louise Brown, so far as we know, is a healthy child, and although it is conceivable that problems may develop later in her life, there is no special reason to believe that will be the case. In any event, if a problem develops later with her, it will be very difficult to trace that back to the way her procreation took place. Hence, at least so far as those moral objections based upon potential to the fetus are concerned, the reasons for worry there seem to have been considerably reduced.

It should be noted in this respect, however, that we have no knowledge of how many failures there were along the line before

Louise Brown, and we can assume that prenatal diagnosis was used to monitor the safety of her development.

What kinds of moral issues does this whole development pose? In one sense, it was an unusual intervention into nature, unusual because it was so different and so new. The motive of the intervention was to satisfy the desires of the parents, rather than to effect anything therapeutic for the child. It is an interesting question whether one wants to call being procreated in the first place a therapeutic improvement for a child. That leads to a number of strange logical puzzles that perhaps we can put aside. Nonetheless, the important thing to bear in mind is that this procedure was used for the sake of the parents and not for the sake of the child. Yet one might say, "Why not?" Infertility is a heavy burden; it is the deprivation of an otherwise normal function for a woman, and for that very reason the use of medical technologies to relieve infertility could be seen, and I believe probably should be seen, as simply the extension of a fairly traditional principle in medicine. That is, use the technologies you have to enable people to achieve what we would call a state of ordinary capacities and possibilities. In the case of in vitro fertilization, a traditional medical function was being performed—to bring a woman who would not otherwise been able to bear a child to a state of normality.

Yet we are still left with some important questions. One might say that, though things turned out well in the case of Louise Brown, a development of such momentous importance should have been decided by society as a whole, and not simply by the individual couple and their interested researcher-physicians. Although the researchers might have been able to make decisions concerning the physical well-being of the developing fetus, they were not in any special position to make a moral judgment concerning the legitimacy of the research in the first place. But was it a matter that ought, then, to have been left exclusively to the parents? My own inclination is to say, yes, it is a moral question appropriate to their choice. There is no reason to believe that society, had it intervened, would have had any particular knowledge or understanding that would have given it a privileged status on such matters. There is no reason either to believe that the researchers have any privileged knowledge, other than technical. Thus, in a situation with that much uncertainty, it would seem appropriate that the parents themselves make the choice.

The difficulty, of course, is that even if one leaves the choice to the parents, then others might still want to know how one assesses the potential consequences of this kind of development. In that respect, the dilemma with in vitro fertilization is a common one, and an increasingly common one, with many technological developments, and perhaps we should see that issue in the broader context of the morality of technological development in general. Given something radically new, there is rarely any theoretical or philosophical way of accurately assessing what the consequences will be. We can surely imagine consequences. We could say that if a certain number of children are born defective as a result of the process, and a certain number are born healthy, then we could have consequences that might be capable of assessment. But at this point, since so few children have been born as a result of the technique of in vitro fertilization, we have no solid knowledge whatever about the long-term, or even the short-term, consequences of the development. We are therefore left with a very difficult choice: we can either say such developments should not go on at all, because there is no way to assess the consequences; or to say that they ought to go on, and that we must simply judge from experience whether the initial choice was a good one.The bias in technological societies, including our own, has been to take the risk, and then to assess the consequences after the fact.

I am myself inclined to think that we ought, prudently, to let things go forward, if only because it is difficult to develop moral principles that seem powerful enough to stop matters prior to any experience whatsoever.

This is not to say that moral analysis should not take place, or that some principles are not pertinent. It is simply to say that in the absence of human experience, in the absence of seeing individual, family, and social consequences, it is not at all clear that the available moral principles apply, or what the principles ought to tell us about making decisions in this area.

In my own case, this represents a change in my thinking. During the earlier debate, in the early 70s, I was one of those who was opposed to in vitro fertilization. I felt that it was an unwarranted hazard to fetuses, and could see no overwhelming reasons why those hazards should be run.

I am willing to say that there were many aspects about the development that were unacceptable, particularly the unwillingness of the researchers themselves to engage in public debate. Nonetheles, since it turned out well—at least so far things have turned out well— I suppose it is now necessary to allow couples to have the possibility of making use of in vitro fertilization. For my part at least, I can think of no strong moral principles now that would allow me to stand in their way. Yet a warning seems in order here also. We know it is very difficult to go back on technologies. Even if it should turn out later there are some very grave consequences in in vitro fertilization, it is more than likely there will be at least some benefits for some people. In that circumstance, it is very difficult to stop and turn back the clock.

One can go down a whole list of technologies over the past few centuries where people did worry about introducing them in the first place. But since there seemed to be no way to set those worries at rest, the tendency has been to move forward with them. In this area— quite new—we have a very short history of dealing with problems of this kind, and I for one am not about to say whether it is likely to prove a good intervention into a natural process or not. I hope so, but I think we should recognize that there are considerable uncertainties, and recognize also that it may take one or more generations to clarify the value of in vitro fertilization and to evaluate its implications for human life.

Chapter 15

Fathers Anonymous

Beyond the Best Interests of the Sperm Donor*

George J. Annas

Alex Haley concludes his international best seller, *Roots*, with the burial of his father in Little Rock, Arkansas. Walking away from the graveside he ponders the past generations, observing "I feel that they *do* watch and guide." The book inspired whole industries devoted to the development of family trees, and locating one's "roots" has become somewhat of an obsession with many. Because of the current secrecy surrounding the practice of artificial insemination by donor (AID), there are an estimated 250,000 children conceived by AID (at the rate of 6,000–10,000 annually in the United States) who will never be able to find their biological roots. There is almost no data available on these children, their psychological development, or their family life. The entire procedure has been shrouded in secrecy that is primarily justified by fear of potential legal consequences should the fact of AID be discovered.

It is the thesis here that most of the informal "policies" concerning AID as it is presently practiced in the United States have

*Copyright 1979, George J. Annas.

come about because of an exaggeration of potential legal pitfalls and a failure to pay sufficient attention to the best interests of the AID child. Accordingly it is at least premature either to legislate "standards" or use AID as a "model" for IVF. Most commentary on AID has concentrated on theoretical legal problems without paying attention to real psychological problems. Indeed, most the legal literature reads like an answer to the following question: "Review all of the case law and statutes relating to AID and discuss all possible lawsuits that any participant or product of AID might have against anyone. If time permits, suggest a statutory scheme that might minimize these problems." Rather than add another answer to this interesting but tangential question, I will review the rationale for AID, the manner in which donors are selected, and the way records are kept, with a view toward developing policies and practices that maximize the best interests of the child.[1]

Why Use AID?

The question of indications is almost never addressed in the medical or legal literature beyond assuming that it is almost exclusively a "treatment for husband infertility." To find a model for AID one must consult the social satirists of the 20th century, and the writings of philosphers. In George Orwell's *1984* reproduction by artificial insemination (although not necessarily by donor) was mandatory as part of a program to remove all pleasure from sexual intercourse. Other measures enacted toward this end were Party approval of all marriages (always refused if the couple was physically attracted toward each other) and promotion of the view that sexual intercourse should be seen "as a slightly disgusting minor operation, like having an enema." All children were raised in public institutions.

Artificial insemination, however, is only one possible consequence, not a cause, of a totalitarian state of the type envisaged by Orwell. In this regard Joseph Fletcher is quite correct in observing that in such a society "the modes of reproduction would be of a relatively minor concern...compared to the many human values certain to be destroyed."[2]

AID is taken more seriously, and viewed with more hope than fear, by Aldous Huxley. In *Brave New World Revisited* he writes that every new advance in medicine will "tend to be offset by a corresponding advance in the survival rate of individuals cursed by some genetic insufficiency...and with the decline of average healthiness there may well go a decline in average intelligence." Huxley presents one solution to this problem in his view of the ideal society, *Island*. In that society AID is no mandatory, but is in fact used by almost everyone—at least for the third child, and by most couples who decide to have only two, for their second child. The rationale is the one previously expounded by Huxley: to increase the general IQ of the population instead of allowing it to gradually decrease. In the words of his character, Vijaya:

> In the early days there were a good many conscientious objectors. But now the advantages of AI have been so clearly demonstrated, most married couples feel that it's more moral to take a shot at having a child of superior quality than to run the risk of slavishly reproducing whatever quirks and defects may happen to run in the husband's family...we have a central bank of superior stocks. Superior stocks of every variety of physique and temperament.[3]

The problems of selecting such "superior stock" have been discussed, but not resolved. H.J. Muller, for example, argued in 1935 that no intelligent and morally sensitive woman would refuse to bear a child of Lenin—while in later version Lenin is omitted and Einstein, Pasteur, Descartes, Da Vinci, and Lincoln are nominated.[4] Theodosius Dobzhansky has noted that "Muller's implied assumption that there is, or can be, *the* ideal human genotype which it is desirable to bestow upon everybody is not only unappealing but almost certainly wrong—it is human diversity that acted as a leaven of creative effort in the past and will so act in the future."[5] This is, of course, simply an axiom of evolution and natural selection. The problem of making conscious choices is that we cannot accurately predict what traits future generations will require for survival.

Sociobiologists have recently identified another genetic "truth" in the animal kingdom that may have relevance to the AID situation in humans, i.e., that animals will try to maximize the spread of their genes. In the words of Richard Dawkins, "Ideally what an individual would 'like' (I don't mean physically enjoy, although he might) would be to copulate with as many members of the opposite sex as possible, leaving the partner in each case to bring up the children."[6]

In this way the genes of the father are distributed maximally. AID and sperm banking remove the previous physical limitations of such a strategy from the human animal.

On the moral plane, AID has been condemned by the Catholic Church (primarily because masturbation is viewed as an unnatural and evil act), and by such writers as Paul Ramsey (on the basis that it is "an exercise of illicit dominion over man").[7] On the other hand, Joseph Fletcher has vigorously defended the morality of AID. He has argued first that there is ample precedent for the practice in the Old Testament (e.g., Deuteronomy 25: 5–6, and Genesis 30: 1–13) and that it is licit means toward a highly desirable end (parenthood for the otherwise sterile couple). His conclusions are based on his belief that fidelity of marriage is a personal bond between husband and wife (not primarily a legal contract), and that parenthood is a moral relationship with children, not a material or merely physical one.[8]

Until early this year it was impossible to even speculate with any authority on the indications for AID in contemporary medical practice. In March of 1979, however, Curie-Cohen, Luttrell, and Shapiro of the University of Wisconsin published their questionnaire survey of AID practitioners. They located 379 practitioners of AID who accounted for approximately 3576 births in 1977 and responded to a series of questions about their practices. The results of this survey provide the only data in existence on the current practice of AID in the United States, and the survey, which will be referred to as the "Curie-Cohen survey," will be cited extensively in this paper.[9]

As to indications, their survey findings are instructive: 95% of the respondents reported that their primary reason for using AID was for husband infertility. However, at least 40% had used AID for other reasons: one-third had used it for fertile couples when the husband feared transmission of a genetic disease (similar to Huxley's *Island* rationale; and almost 10% had used it to fertilize single women (removing sex from reproduction altogether, and highlighting fears of many moralists). Therefore, whatever one views as society's rationale for permitting AID to continue, it must be recognized that a large percentage of practitioners are using it for eugenic purposes. In addition, those that use it to fertilize single women or members of lesbian couples are engaged in a practice that most of society would probably condemn because of its implications for the child and the family as a basic unit of society.

The issue of indications needs to be faced directly and clearly by commentators and practitioners alike so that an informed consensus can be reached. It is worth observing, however, that current rationales for servicing the infertile couple, the lesbian couple, and single women all rest primarily on one's definition of the best interests of the couple or prospective parent, and not on the best interests of the child. Although many physicians "screen" recipients to determine whether their motives are proper and their marriage "stable," there is no evidence that they are competent to make these judgments. This is not necessarily to say that AID should not be available to couples in which the husband is sterile; it is only to highlight the fact that we have no data concerning how children born into this situation fare, and to suggest that it is irresponsible to continue the practice of AID for this indication without attempting to gather such data.

Donor Selection

Donor selection may be the most difficult issue in AID, but legal considerations are not controlling. First it should be noted that "donor" is a misnomer. Virtually all respondents in the Curie-Cohen study paid for ejaculates, 90% paying from $20 to $35 per ejaculate, with 7% paying more, up to $100. Thus a more accurate term would be "sperm vendors." Although this distinction may seem trivial, it has legal consequences. For example, it makes no sense to designate the form signed by the "vendor" as a "consent form" since he is not a patient and is not really consenting to anything. It is a contract in which the sperm vendor agrees to deliver a product for pay. We can debate the elements of the agreement, but must would probably agree that it should spell out the vendor's obligations in terms of his own physical and genetic health, including an accurate family history, the quality of the specimens he is required to produce, the necessity for complete and permanent anonymity of the recipient, and a waiver of any rights in any child resulting from the insemination. In return, the buyer agrees to pay the vendor and protect his anonymity.

The issue of *who* selects the sperm vendor has been given far too little attention. The Curie-Cohen study found two things of interest in this regard. First, 92% of practitioners never permit the recipient to select the donor, although the remainder do on rare occasion. Also, 15% used frozen semen obtained from sperm banks, and others used sperm from those selected by urologists or other personal associates. The point is that at least in a small minority of cases, someone other than the physician selects the source of the sperm. More significant, however, is the fact that almost all physicians make their own selection, most using medical students. Sixty-two percent used medical students or hospital residents; 10% used other university or graduate students; 18% used both, and the remaining 10% used donors from military academies, husbands of obstetric patients, hospital personnel, and friends.

Physicians in all of these situations are making eugenic decisions —selecting what they consider "superior" genes for AID. In general they have chosen to reproduce themselves (or those in their profession), and this is what sociobiologists like Dawkins would probably have predicted. Although this should not surprise us, it should be a cause of concern, since what may be controlling is more than just convenience. Physicians may believe that society needs more individuals with the attributes of physicians, but it is unlikely that society as a whole does. Lawyers would be likely to select law students; geneticists, graduate students in genetics; military personnel, students at the military academies, and so on. The point is not trivial. Courts have found in other contexts that physicians have neither the training nor the social warrant to make "quality of life" decisions. In the *Houle* case, for example, a physician's decision not to treat a defective newborn was overruled on the basis that "the doctor's qualitative evaluation of the value of the life to be preserved is not legally within the scope of his expertise." Selecting donors in this manner, rather than matching for characteristics of the husband, for example, seems to be primarily in the best interest of the physician rather than the child, and can probably not be justified. Nor can the argument that medical students know more about genetics than other graduate students stand analysis. They are probably also just as susceptible to monetary influence as are some of the blood sellers described in Richard Tittmuss' classic study, *The Gift Relationship*. Perhaps national guidelines, developed by a committee made up of a random sample of the population, would be more appropriate.

The Curie-Cohen survey also revealed that even on the basis of simple genetics, physicians administering AID "were not trained for the task" and made many erroneous and inconsistent decisions. Specifically, 80–95% of all respondents said they would reject a donor if he had one of the following traits, and more than 50% of all respondents would reject the same donor if one of these traits appeared in his immediate family: Tay-Sachs, hemophilia, cystic fibrosis, mental retardation, Huntington's chorea, translocation or trisomy, diabetes, sickle-cell trait, and alkaptonuria. This list includes autosomal recessive diseases in which carriers can be identified, and those in which they cannot, dominant, X-linked, and multigenic diseases.

The troubling findings are that the severity and genetic risk of the condition was not reflected in rejection criteria, and that genetic knowledge appears deficient. For example, 71% would reject a donor who had hemophilia in his family, even though this X-linked gene could *not* be transmitted unless the donor himself was affected. Additionally, although 92% said they would reject a donor with a translocation trisomy, only 12.5% actually examined the donor's karyotype. Similarly, while 95% would reject a carrier of Tay-Sachs, fewer than 1% actually tested donors for this carrier state. In fact, only 29% performed any biochemical tests on donors other than blood typing, and these tests were primarily for communicable diseases. The conclusion must be that while prevention of genetic disease is a goal, it cannot be accomplished by the means currently in use. The findings also raise serious questions about the ability of these physicians to act as genetic counselors, and suggest that other nonmedical professionals may be able to do a better job in delivering AID services in a manner best calculated to maximize the interests of the child.

Since there is a most uniform agreement that certain genetic conditions contraindicate use of a person's sperm for AID, it is likely that a court would find a physician negligent in using such sperm even though few physicians actually test to make sure the sperm vendor is not affected.[10] "There are precautions so imperative that even their universal disregard will not excuse their omission."[11] This is an area in which uniform standards need to be developed within the profession.

The other related issues concerning the donor or sperm vendor merit mention because they have apparently been dealt with strictly on the basis of fear of legal liability rather than any social or medical rationale or concern for the best interests of the child: consent of the donor's wife and record keeping.

The Donor's Wife

The American Medical Association, the British Medical Society, and authorities in Australia all argue, as do almost all legal commentators, that the wife of the sperm donor must sign the "consent" form "because marital interests are involved." None of these sources or commentators, however, provide any further explanation. This type of advice can be viewed as a paradigm of legalism based on fear and ignorance.

I do not know what the original source of this recommendation is, but it may be Joseph Fletcher's comments in 1954: "...it is clearly a requirement of personal integrity, of love and loyalty, that the donor's wife should be consulted by him (the donor) and agree to the role he plays."[12] Perhaps. But however one comes out on this pro-nouncement, it is not a legal requirement, and does not seem to serve any useful social purpose. In terms of liability on the part of the physician, the potential grounds appear to be two: (1) an action in contract to recover a portion of the money received by the husband for his sperm on the grounds that the wife has a property interest in her husband's sperm; and (2) an action for alienation of affections by the wife against the physician on the basis that her husband prefers masturbation for pay to intercourse with her, or some other fantasy he may have developed that interferes with the marriage. Both of these strike me as being too silly to worry about, and any woman who would bring either action is not likely to be discouraged by the fact that she has signed a "consent" (read contract) form. In addition, such a requirement is at odds with more recent United States Supreme Court decisions that refuse to permit one spouse to have veto power over procreation decisions made by the other spouse. Specifically, a husband may not be required to consent to his wife's abortion by state law because her right to make this decision is constitutionally protected.[13]

Record-Keeping

Though the Curie-Cohen survey found that 93% of physicians kept permanent records on recipients, only 37% kept permanent records on children born after AID (fewer than the 50% who provided obstetric care for their inseminated patients), and only 30% kept any permanent records on donors. Moreover, 83% opposed any legislation that would mandate the keeping of records because it would make protection of anonymity of the donor more difficult. The fear of record-keeping seems to be based primarily on the idea, common in the legal literature, that if indentifiable, the donor might be sued for parental obligations (e.g., child support, inheritance, and so on) by one of his "biological children" sired by the AID process, and that this suit might be successful. The underlying rationale is that without anonymity assured, there would be no donors. There are a number of responses to this argument:

1. It is important to maintain careful records to see how the sperm "works" in terms of outcome of the pregnancy. If a donor is used more than once, a defective child should be grounds for immediately discontinuing the use of the sperm for the protection of potential future children. Since the survey disclosed that most

physicians have no policy on how many times they use a donor, and 6% had used one for more than 15—with one using a donor for 50 pregnacies—this issue is much more likely to affect the life of a real child than the highly speculative lawsuit is to affect a donor.

2. No meaningful study of the characteristics of donors can ever be made if there are no records kept concerning them.

3. In those cases where family history is important (and it is important enough to ask *every* donor about his) the AID child will *never* be able to respond accurately.

4. Finally, and most importantly, if no records are kept, the child will *never*, under any circumstances, be able to determine its genetic father. Since we do not know what the consequences of this will be, it cannot be said that destroying this information is in the best interests of the child. The most that can be said for such a policy is that it is in the best interests of the donor. But this is simply not good enough. The donor has a choice in the matter, the child has none. The donor and physician can take steps to guard their own best interests, the child cannot.

Given the recent history of adopted children, it is likely that if AID children learn they are the products of AID, they will want to be able to identify their genetic father. It is now relatively accepted practice to tell adopted children that they are adopted as soon as possible, and make sure they understand it. This is because it is thought they will inevitably find out some day, and the blow will be a severe one if they have been lied to. In AID, the concensus seems to be not to tell on the basis that no one is ever likely to find out the truth, since to all the world it appears that the pregnancy proceeded in the normal course.

Moralists would probably agree with Fletcher that the physician should not accept the suggestion that a husband's brother be used as a donor without the wife's knowledge (his intent is to keep the blood line in his children) because this is a violation of "marital confidence." It seems to me a similar argument can be made of consistently lying to the child—i.e., that it is a violation of parental–child confidence. There is evidence that AID children do learn the truth, and the only thing all fifteen states with legislation on AID agree to is that it should legitimize the child—an issue that will never arise unless the child's AID status is discovered. If AID is seen as a loving act for the child's benefit, there seems no reason to taint the procedure with a lie that could prove extremely destructive to the child.

A number of policies would have to be changed to permit open disclosure of genetic parenthood to children. The first is relatively easy: a statute could be enacted requiring the registration of all AID children in a court in a sealed record that would only be available to the child; the remainder of the statute would provide that the genetic father had no legal or financial rights or responsibilities to the child. A variation on this would be to keep the record sealed until the death of the donor, or until he waived his right to privacy in this matter. In the long term, a more practical solution may lie in only using the frozen sperm of deceased donors. In this case full disclosure could be made without any possibility of personal or financial demands on the genetic father by the child.[14]

Worry about donors, in any event, is probably out of proportion to reality. There have been *no* suits against any donor by any child even though almost one-third of physicians engaging in AID keep permanent records of the donors. No matter what steps are taken to protect them, it seems essential to me that, in the potential best interests of the child, such records be kept and that their contents be based on the development of professional standards for such records.

Not keeping records can also lead to other bizarre practices. For example, some physicians use multiple donors in a single cycle to obscure the identity of the genetic father. The Curie-Cohen survey found that 32% of all physicians utilize this technique, which could be to the physical detriment of the child (and potential future of a donor with defective sperm), and cannot be justified on any genetic grounds whatsoever.[15]

Summary and Conclusions

Current AID practices are based primarily on consideration of protecting the interest of practitioners and donors rather than recipients and children. The most likely reason for this is found in exaggerated fears of legal pitfalls. It is suggested that policy in this area should be dictated by maximizing the best interest of the resulting children. The evidence from the Curie-Cohen survey is that current practices are dangerous to children and must be modified. Specifically, consideration should be given to the following:

1. Removing AID from the practice of medicine and placing it in the hands of genetic counselors or other nonmedical

personnel (alternatively, a routine genetic consultation could be added for each couple who request AID).

2. Development of uniform standards for donor selection, including national screening criteria.

3. A requirement that practitioners of AID keep permanent records on all donors that they can match with recipients; I would prefer this to become common practice in the profession, but legislation requiring filing with a governmental agency may be necessary.

4. As a corollary, mixing of sperm would be an unacceptable practice; and the number of pregnancies per donor would be limited.

5. Establishment of national standards regarding AID by professional organizations with input from the public.

6. Research on the psychological development of children who have been conceived by AID and their families.

Dr. S.J. Behrman concludes his editorial on the Curie-Cohen survey by questioning the "uneven and evasive" attitude of the law in regard of AID, and recommending immediate legislative action:

> The time has come—in fact, is long overdue—when legislature must set standards for artificial insemination by donors, declare the legitimacy of the children, and protect the liability of all directly involved with this procedure. A better public policy on this question is clearly needed.[15]

I have suggested that agreement with the need for "a better public policy" is not synonymous with immediate legislation. The problem with AID is that there are many unresolved problems with AID, and few of them are legal. There is no social or professional agreement on indications, selection of donors, screening of donors, mixing of donor sperm, or keeping records on sperm donations. Where there is agreement, such as in requiring the signature of the donor's wife on a "consent" form, the reasons for such agreement are unclear.

It is time to stop thinking about uniform legislation and start thinking about the development of professional standards. Obsessive concern with self-protection must give way to concern for the child.[16]

Notes

1. Currently at least fifteen states (Alaska, Arkansas, California, Florida, Georgia, Kansas, Louisiana, Maryland, New York, North Carolina, Oklahoma, Oregon, Texas, Virginia, and Washington) have statutes on the books that mention AID. All of these statutes specifically provide that the resulting child is the natural child of the recipient's husband provided he has consented to the procedure. Five states require that the consent be filed with a state agency (Kansas, Oklahoma, Georgia, Washington, and Oregon) and six states, either directly or by implication, limit the practice of AID to physicians (California, Oklahoma, Virginia, Washington, Alaska, and Oregon). Only two states, Washington and Texas, specifically provide that the sperm donor is not the father of the child. Oregon's is the only criminal statute, and makes it a Class C misdemeanor, punishable by 30 days in jail, for anyone but a physician to select sperm donors, and for a donor to provide semen if he "(1) has any disease or defect known to him to be transmissible by genes; or (2) knows or has reason to know he has a venereal disease." The only state supreme court to ever rule on AID held a consenting husband liable for child support. People v. Sorensen, 437 P.2d 495 (Cal. 1968). An excellent overview of the law, which will not be repeated in this article, appears in the proceedings from the first conference, Milunsky, A., and Annas, G. J. *Genetics and the Law*, Plenum: New York, 1976; Healey, J. *Legal Aspects of Artificial Insemination by Donor and Paternity Testing*, 203-218.
2. Fletcher, J. *Morals and Medicine*, Beacon Press: Boston, 1954, 134.
3. Huxley, A. *Island*, Perennial Classic: New York, 1972, 193-194.
4. Quoted by Ramsey, P., *Fabricated Man*, Yale U.: New Haven, 1970, 49.
5. *Id.* at 53.
6. Dawkins, R. *The Selfish Gene*, Oxford University Press: New York, 1976, 151.
7. Ramsey, P. *Fabricated Man*, *supra.* note 4 at 48.
8. Fletcher, J. *Morals and Medicine*, *supra.* note 2 at 116–122.
9. Curie-Cohen, M., Luttrell, L., and Shapiro, S. Current practice of artificial insemination by donor in the United States, *New Engl. J. Med.* **300**: 585, 1979.
10. While there is no specific legal standard for screening sperm donors, when done by a physician the general law of specialists is is applicable:

    One holding himself out as a specialist should be held to the standard of care and skill of the average member of the profession practicing in the specialty, taking into account the advances in the profession.

A recent analogous case involved an individual who received two cornea transplants. The transplanted corneas turned out to be infected, and caused total and pemanent blindness in the recipient. He sued the hospital and the resident who had removed the donor's eyes. The jury found in favor of the resident, but against the hospital. In affirming the jury's verdict against the hospital, the court noted that although the hospital had "no printed or published checklist which could be used as a guide-line for determining the suitability of a prospective donor" there was testimony that published criteria did exist and were "fairly uniform throughout the nation." The court further concluded that had these criteria been applied to the donor in this case, the jury could have rightfully decided that he would have been rejected:

The jury heard expert testimony to the effect that cadavers with a history of certain types of illnesses are not generally wise choices for cornea donation. It follows that whoever may have had the responsibility of determining the suitability of the cornea for transplant would have been *required*, in the exercise of due care, *to review carefully and exhaustively the medical history of the proposed donor*...The jury could have determined that Detroit General was negligent in failing to set up a procedure which would assure that the party responsible for determining the suitability of the cornea for transplant would have *access to all the relevant medical records of the proposed donor*. (emphasis supplied)

Applied to AID donors, this case indicates that hospitals and physicians are responsible for determining the suitability of donors and, if they do not have a reasonable policy of their own, will be held to whatever policy has been accepted by other professionals engaged in the same activity.

11. Hooper, T. J. 60 F. 2d 737, 740 (2d Cir. 1932); and see Helling v. Carey, 519 P. 2d 981 (Wash. 1974).
12. Fletcher, J. *Morals and Medicine*, *supra.*, note 2, 129.
13. For a fuller discussion of this issue see Glantz, Leonard. *Recent Developments in Abortion Law* in Milunsky, A., and Annas, G. J., editors, *Genetics and Law II*, Plenum: N.Y., 1980.
14. Sperm banks may soon begin marketing sperm directly to consumers, bypassing physicians and adding to current confusion in practice. See *Advertising Age,* May 14, 1979 at 30.
15. A more encouraging finding was that only two physicians in the entire sample mixed donor sperm with the husband's semen. This apparently once common practice has died, probably because

it is now known to be medically contraindicated. See Quinlivan, W. L. G., Sullivan, H.: Spermatozoal Antibodies in Human Seminal Plasma as a Cause of Failed Artificial Donor Insemination. *Fertil Steril.* **28**: 1082–1085, 1977.

16. Behrman, S. J. Artificial Insemination and Public Policy, *New Engl. J. Med.* **300**: 619–620, 1979.

17. The argument is not based on any alleged action for "wrongful life" that the child may have, but on the theory that we should do what we can to protect the interests of "innocent" third parties whenever their interests are in conflict with those who have the ability to affect them.

Chapter 16

Law and Ethics

Three Persistent Myths

F. Patrick Hubbard

Introduction

The papers of Drs. Callahan and Annas have proved both enlightening and thought-provoking. I find myself wanting to raise a dozen or more different points of fundamental agreement or disagreement.

In this commentary, however, I will confine myself to one central concern. What I hope to do is develop a background for discussing the interrelationships of law, technology (including, of course, reproductive technology and "heroic" life-saving techniques), and ethics. This background is necessary because of three persuasive misunderstandings or myths about the nature of law. These myths often result in confusion concerning specific issues, such as, for example, the obligations of a doctor in using artificial insemination.

The Myth That Law Is All-Inclusive

The first misconception is that law is all-inclusive—i.e., that given some particular type of conduct, there is *a* discrete rule or regulation that is applicable to this conduct. There is simply no such coverage. Life is too complex for any system of rules to encompass all the questions that arise. This situation is particularly true where technology is involved because lawmakers cannot possibly anticipate all the technological developments that will present new and unexpected isssues.

For a variety of reasons, many people find it very bothersome to view law as so incomplete that numerous social acts take place in a context that is "nonlegal." To these persons, conduct—whether it involves using "heroic" life–saving measures or reproductive technology—must be either legal or illegal. The notion that it could be neither legal nor illegal seems absurd.

The solution to this dilemma is to divide law into two categories of rules. The first type of law tells a citizen—for example, a physician or a medical researcher—what is forbidden, what is required, and what is permitted. This first type of law will be incomplete, however, because of the complexities discussed above. This incompleteness is taken care of by the second type of law, which authorizes officials to fill the gaps resulting from the incompleteness of the first type of law. From the perspective of this two-category system law there is perhaps a sense in which law can be said to be all-inclusive. Whether we are satisfied with this "solution" though is another matter, one that can be addressed after considering the second myth about law.

The Myth That Law is Determinate or Certain

This second misconception is that law is determinate or certain. By this I mean that it makes sense to speak about "what the law *is*" rather than in terms of "our guesses about what the law *might be*." The Annas paper, I feel, frequently encounters difficulties because of its reliance on the assumption that it is meaningful to speak about "what the law is."

This assumption is invalid for a number of reasons, but I think that it is only necessary to discuss two of them. First, it is often based on the view that law is all-inclusive, yet the preceding discussion indicates how incomplete law is. Dividing rules into two types does not help us here. Even if a judge is authorized to fill the gaps in the first type of rules, we do not know for certain how he or she will fill that gap until after has been filled.

The invalidity of the assumption that one can speak with certainty about legal requirements results also from the vagueness and ambiguity inherent in the documentary sources of rules. There is a tendency to think that a judicial precedent or a statute that addresses an issue settles all the legal aspects of that issue. For a surprisingly small category of "clear cases" this is so. But for a very large number of cases it is not. For every accepted conception of the proper approach to reading a statute or case, there is a diametrically opposed, but equally respected view that could result in an opposite interpretation. The recent case of *United Steelworkers v. Weber*, 99 S. CT. 2721 (1979) illustrates this tension. The majority, following an accepted practice of looking to the purpose of a statute to interpret its meaning, found that the Civil Rights Act did not forbid a private employer from utilizing a training program that favored blacks. The dissenters, using another traditional scheme of statutory interpretation, found that the "plain meaning" of the words in the act was that the discriminatory training program was forbidden. Statutes and cases simply do not provide certain guidance.

Perhaps the point about uncertainty can be made in another way. Let us contrast three counties in a mythical state. (Let's call it "Petigru.") Petigru has a general statute prohibiting murder, and case authority clearly indicates that actively assisting a person to commit suicide is murder. In two counties in the state every "suspicious" death in hospitals is investigated and indictments of doctors and nurses for murder of terminal patients in great pain are common. In one of these counties, two doctors have been convicted in the past five years. In the other, no one has ever been convicted despite clear evidence suggesting active assistance in a suicide. In the third county, the police never investigate hospital deaths and the public prosecutors never seek indictments concerning such deaths. This situation of three different approaches (which is not uncommon in our pluralistic, federal form of government) raises a "bottom-line" question: What is the law of Petigru? There is no clear, certain answer to this question.

The Myth That Law Tells Us What We Ought to Do

Now I would like to address the third frequent misconception about law. This is the idea that law tells us what we ought to do. From this perspective, if something is legal, then it is alright; if an act is illegal, then it is wrong. There is, of course, a considerable amount of truth in this, but it is a serious error to mistake this partial truth for the whole story.

Before I address the reasons why this view of law is at least a partial misconception, let me stress the core of good sense in the view that the law is a guide to proper behavior. In any resonably well-ordered society it seems fairly obvious that citizens should have at least a prima facie obligation to obey the law. Since I believe we live in such a society, I also believe, for example, that a doctor has both a legal and moral or ethical obligation not to perform an act of active euthanasia. Nevertheless, this duty is not absolute in a moral sense; it may be that a doctor under some compelling circumstance might feel a conflicting and overwhelming moral obligation to break the law and assist a person who wants to die, but who is unable to commit suicide without active assistance. But let me repeat: Absent such a compelling situation, there is a moral obligation to obey the law.

There are also prudential reasons for obeying the law. Apart from any internal feelings of obligation to respect legal prohibitions, active euthanasia is not a wise course of action for the prudent doctor to follow. Intentional homicide is murder (or at the very least volun-tary manslaughter) and is punished severely in our society.

At this point I would like to pause a moment to elaborate on prudential reasons for obeying the law and to develop my earlier point that Annas is mistaken when speaking about "what the law *is*" rather than "our predictions of what it *may be*." Uncertainty about legal requirements charaterizes most of the area discussed by both Callahan and Annas. Yet Annas has implied by his remarks that there is something wrong or unethical involved when doctors and their lawyers elect to be prudent about the risks in this area. He claims that they are using the law as a "smoke screen" because the law is not what they claim it "is." But to repeat: An attorney advising a doctor cannot often speak about what the law "is" in questionable areas of bioethics. The attorney can only make guesses and try to help the client make choices that are legal, prudent, and ethical. Thus, when a doctor says that he or she will not refrain from life-saving measures unless there is a court order authorizing such res-traint, that physician is not necessarily saying that the law "*is*" that such orders are required, but only that there is a good chance that such an order *might* be required, and that it is prudent to follow this possible interpretation of the law. This does not strike me as unpro-fessional or as "hiding" behind the law. Thus, when we speak of "obeying the law," we often mean prudently arranging our affairs in terms of what the law might be.

Let us return now to my point that, despite the good sense in viewing law as a partial guide to proper behavior, it is a mistake to view legal requirements as equivalent to ethical requirements. One reason why it is a mistake to expect such an equivalence is that there are so many "gaps" in the law. As indicated in discussing the myth of all-inclusiveness, the legal system simply does not address many issues and as a result a particular course of conduct might be neither legal or illegal.

Another reason why legal requirements do not necessarily indicate what actions ought to be taken is that law has many functions other than providing such guidance. For example, one purpose of law is to promote social stability. If people are sufficiently upset by some technique of reproduction—"test tube babies," for example—the technique may be prohibited for that reason alone. Morality could have nothing to do with the decision.

Another way to promote stability is to avoid decisions that explicitly indicate that one fundamental value has yielded to another. For example, contraception was illegal for years in most states, yet it was widely practiced. In this way society could "have its cake (in this case, a formal affirmation of the immorality of contraception) and eat it too"—i.e., be free to decide as individuals whether to use contraception. Another example of conflict in basic values is the clash between saving lives versus reducing suffering. Once again the legal system works very hard to devise a way to avoid an explicit, all-or-nothing resolution of the conflict. In this case, the jury is perhaps the clearest example of an avoidance mechanism. When the jury determines guilt or innocence in a euthanasia case, there is nothing to prevent them from striking a balance in favor of preventing suffering by acquitting the defendant. The legal system is perhaps less concerned with this "bending" of the laws against taking life than with preventing the jury members from ever telling us exactly what they did. Morality, in short, is less important than avoiding an explicit decision that would reveal basic value conflicts and thus threaten the goal of social stability.

Another goal of law in our society is to provide a framework within which autonomous persons, acting individually or co-operatively, can formulate and implement their own life plans. In a very real sense, it is not the purpose of law to tell us what to do so long as we observe certain minimal duties like avoiding intentional harm. It is then up to us to work our arrangements privately.

This approach is, I think, involved to some extent in the AID situation discussed by Annas. He asserts that doctors are not legally required to protect sperm "donors" (or "sellers" if you prefer). Although agreeing to a considerable extent with this assertion about doctors' duties, I draw very different conclusions from it from Annas. He believes that this lack of specific requirements is a "problem" that must be solved either by explicit legal rules or by a clear code of professional conduct. I am somewhat puzzled why he feels that professional standards are the proper solution here since the development of these standards would be plagued by the same unresolved policy questions that he believes will hamper the drafting of legislation. Is it, for example, so absolutely clear that the doctor owes the sperm donor/seller *no* obligations whatsoever?

My disagreements with Annas are more fundamental than this though. He concludes that the lack of legal guidance (or explicit professional standards) here is a *problem*. On the other hand, this lack of explicit legal direction appears to me to be merely part of the legal system's development of a framework for private individuals to order their own lives. Consequently, I am not so sure that there is a problem here at all. If prospective AID parents want additional protection for themselves and their child, they are free to bargain with doctors or donors/sellers for such protection. They could, for example, offer to pay double the normal rate for the donor/seller sperm if protections such as those recommended by Annas are utilized.

This laissez-faire approach is often rejected today. One reason stated in support of such a rejection is that people do not know enough about the risks involved; the law, therefore, must protect them. Thus, a person could urge legislation concerning artificial insemination (or any other reproductive technology) on the ground that the parents, being unfamiliar with or unappreciative of the risks involved, would not be able to bargain effectively with doctors and donors. Arguably the best way to protect the parents and to serve the goal of autonomy here would be to educate potential parents concerning the relevant risks or to impose duties of disclosure on the doctors. Such limited measures, however, would not blunt the other two common criticisms of limited governmental interference in decisions.

The first such objection to the laissez-faire approach is that it is based not only upon people's willingness to bargain for particular technologies such as artificial insemination, kidney machines, or abortion, but also upon their ability to pay for them. If a person feels that the existing distribution of goods in society is unfair, he or she is likely to choose imposed requirements to insure a fair distribution of medical technology. Incidentally, the distribution of the benefits of the techniques discussed by Callahan is an important issue that I think he should have developed.

I would also have preferred to see Callahan explore another problem with a laissez-faire approach in the health care area. This is that many persons are not able to make decisions—the unconscious person (Karen Quinlan, for example). This difficulty is most apparent in the reproductive field discussed by Callahan since it is impossible for unborn persons to bargain concerning their welfare.

Because of my desire to explicate what Annas might mean by his assertion that the lack of specific requirements in AID is a "problem," I have perhaps gone on too long in developing this point about the provision of a framework for cooperative bargains as being one of the conflicting purposes of law. So let me summarize: One reason that law is not a clear guide to ethical conduct is that it has purposes other than providing such a guide—for example, promoting social stability and providing a framework for persons to make their own moral decisions.

One final reason for skepticism about the equivalence of legal requirements and ethical strictures is that decisions by the legal system are all too frequently wrong. Controlling technology is a complicated task and, as the recent events in nuclear power indicate, serious miscalculations about health and safety are very possible.

Conclusions

I have here been focusing on developing a general understanding about law, rather than offering very many specific comments about the two papers. I have done this for two basic reasons.

The first is that I find it difficult to comment on these papers absent a framework such as that I have developed today. I hope that my references to the papers illustrate the utility of this framework.

The second reason is that many readers will in time have occasion to consult lawyers about issues such as those raised here. Perhaps I should be more blunt: Whenever any reader faces a decision such as many of those discussed in this volume, then a lawyer should definitely be consulted. It would be folly to do otherwise because of the prudential concerns about avoiding legal difficulties that I mentioned earlier. My hope is that when such legal advice is sought, the discussions with the lawyer will be more meaningful because there is a clear understanding that law is not all-inclusive, not certain, and not a clear guide to ethical conduct. With this understanding the lawyer's role may be better appreciated as that of an advisor who tries to predict what the law might be and helps in making prudent decisions. The lawyer may, of course, discuss ethics with you, but his or her expertise is not in this area; ultimately, any lawyer will tell you that the final decision is yours.

Further Readings

Suggested Readings for Part I

Baier, Kurt. *The Moral Point of View: A Rational Basis of Ethics.* New York: Random House, 1968.

"Bioethics and Social Responsibility." *Monist* **60** (January, 1977). Special issue.

Branson, Roy. "Bioethics as Individual and Social: The Scope of a Consulting Profession and Academic Discipline." *Journal of Religious Ethics* **3** (Spring, 1975): 111–139.

Callahan, Daniel. "Bioethics as a Discipline." *Hastings Center Studies* **1** (No. 1, 1973), 66–73.

Davis, A., and Aroskar, M. *Ethical Dilemmas In Nursing Practice.* New York: Appleton-Century-Crofts, 1978.

Fenner, Kathleen. *Ethics and Law in Nursing.* New York: D. Van Nostrand, 1980.

Fox, Renee C. "Advanced Medical Technology–Social and Ethical Implications." In Inkeles, Alex et al., eds., *Annual Review of Sociology* **2**, Palo Alto, Calif: Annual Reviews, 1976, p. 231–268.

Frankena, William K. *Ethics.* 2nd ed., Englewood Cliffs, NJ: Prentice-Hall, 1973.

Kant, Immanuel. *Foundations of the Metaphysics of Morals: Text and Critical Essays.* Ed., Robert P. Wolff. New York: Bobbs-Merrill, 1969. (Originally published 1785).

Mill, John Stuart. *Utilitarianism and Other Writings.* Cleveland: Meridian, 1962. (Originally published 1863).

Morison, Robert S. "Rights and Responsibility: Redressing the Uneasy Balance." *Hastings Center Report* **4** (April, 1974).

Pellegrino, Edmund, "Medicine, History and the Idea of Man." *Annals of the American Academy of Political and Social Science* **346** (March, 1963): 9–20.

Ross, W. David. *Foundations of Ethics.* New York: Oxford University Press, 1939.

Williams, Bernard. *Morality: An Introduction to Ethics.* New York: Harper and Row, 1972.

Suggested Readings for Part 2

Annas, George J. *The Rights of Hospital Patients*. An American Civil Liberties Union Handbook. New York: Avon Books, 1975.

Ayd, Frank J., ed., *Medical, Moral and Legal Issues in Mental Health Care*. Baltimore: Williams and Wilkins, 1974.

Birnbaum, Morton. "The Right to Treatment." *American Bar Association Journal* 46 (May 1960): 499–505.

Canterbury v. Spencer, 464 F 2d 722, 791 (DC Cir. 1972).

"Developments in the Law-Civil Commitment of the Mentally Ill." *Harvard Law Review* 87 (April 1974): 1190–1406.

Gaylin, Willard. "What's Normal?" *New York Times Magazine*, April 1, 1973: pp. 14ff.

Hartmann, H. *Psychoanalysis and Moral Values*. New York: International Universities Press, 1960.

Lazare, Aaron. "Hidden Conceptual Models in Clinical Psychiatry." *New England Journal of Medicine* 288 (Feb. 15, 1973): 345–351.

Murphy, Jeffrie G. "Total Institutions and the Possibility of Consent to Organic Therapies." *Human Rights* 5 (Fall, 1975): 25–45.

Robitscher, Jonas. "The Right to Psychiatric Treatment: A Social–Legal Approach to the Plight of the State Hospital Patient." *Villanova Law Review* 18 (November 1972): 11–36.

Spece, Roy G., Jr. "Conditioning and Other Technologies used to 'Treat?' 'Rehabilitate?' 'Demolish?' Prisoners and Mental Patients." *Southern California Law Review* 45 (Spring 1972): 616–684.

Stone, Alan. "Overview: The Right to Treatment–Comments on the Law and Its Impact." *American Journal of Psychiatry* 132 (Nov. 1975): 1125–1134.

Szasz, Thomas. *Ideology and Insanity*. New York: Doubleday Anchor, 1980.

—. *Manufacture of Madness: A Comparative Study of the Inquisition and the Mental Health Movement*. New York: Harper and Row, 1970.

—. *Myth of Mental Illness*, New York: Hoeber, 1961.

Suggested Readings for Part 3

Behnke, John A., and Bok, Sissela. *The Dilemma of Euthanasia*. Garden City, New York: Doubleday Anchor Books, 1975.

Bok, Sissela. "Personal Directions for Care at the End of Life." *New England Journal of Medicine* 295 (August 12, 1976): 367–369.

Branson, Roy, "Is Acceptance a Denial of Death? Another Look at Kübler-Ross." *Christian Century*, May 7, 1975: 464–468.

Brill, Howard W. "Death with Dignity: A Recommendation for Statutory Change." *University of Florida Law Review* 12 (Winter 1970): 368–383.

Cantor, Norman L. "A Patient's Decision to Decline Life-Saving Medical Treatment: Bodily Integrity Versus the Preservation of Life." *Rutgers Law Review* **26** (Winter 1972): 228–264.

Downing, A. B. *Euthanasia and the Right to Die.* New York: Humanities Press, 1970.

Dyck, Arthur. "An Alternative to the Ethic of Euthanasia." In Williams, Robert H., ed., *To Live and To Die: When, Why, and How.* New York: Springer-Verlag, 1973, p. 98–112.

Engelhardt, H. Tristam. "Euthanasia and Children: The Injury of Continued Existence." *Journal of Pediatrics* **83** (July 1973): 170–171.

Fletcher, George. "Prolonging Life." *Washington Law Review* **42**, (1967): 999–1016.

Fletcher, Joseph. "Ethics and Euthanasia." In Williams, Robert H., ed. *To Live and To Die: When, Why, and How.* New York: Springer-Verlag, 1973, p. 113–122.

Foot, Philippa. "Euthanasia." *Philosophy and Public Affairs* **6** (Winter 1977): 85–112.

Group for the Advancement of Psychiatry. *The Right to Die: Decision and Decision Makers.* Proceedings of a Symposium, vol. VIII, symposium No. 12, November 1973. New York, 1973.

Gustafson, James M. "Mongolism, Parental Desires, and the Right to Life." *Perspectives in Biology and Medicine* **16** (Summer 1973), 529–557.

Hare, R. M. "Euthanasia: A Christian View." *Philosophic Exchange* **2** (Summer 1975): 43–52.

Kohl, Marvin, ed., *Beneficent Euthanasia.* Buffalo: Prometheus Books, 1975.

Kubler-Ross, Elisabeth. *On Death and Dying.* New York: Macmillan, 1969.

Maguire, Daniel C. *Death By Choice.* Garden City, New York: Doubleday 1974.

McCormick, Richard A. "To Save or Let Die." *Journal of the American Medical Association* **229** (July 8, 1974): 172–176.

Massachussetts General Hospital, Clinical Care Committee. "Optimum Care for Hopelessly Ill Patients." *New England Journal of Medicine* **295** (August 12, 1976): 362–364.

Paris, John J. "Compulsory Medical Treatment and Religious Freedom: Whose Law Shall Prevail?" *University of San Francisco Law Review* **10** (1975): 1–35.

Shaw, Anthony. "Dilemmas of 'Informed Consent' in Children." *New England Journal of Medicine* **289** (October 25, 1973): 885–890.

Veatch, Robert M. *Death, Dying and the Biological Revolution.* New Haven: Yale University Press, 1976.

Williams, Glanville. "Euthanasia and Abortion." *University of Colorado Law Review 38* (1966).

Suggested Readings for Part 4

Bergsma, Daniel, ed. *Ethical, Social, and Legal Dimensions of Screening for Human Genetic Disease.* Birth Defects: Original Article Series, Vol. 10, No. 6. Miami: Symposium Specialists, 1974.

Callahan, Daniel. "What Obligations Do We Have to Future Generations?" *American Ecclesiastical Review* **164** (April 1971): 265–280.

Edwards, Robert G. "Fertilization of Human Eggs in Vitro: Morals, Ethics, and the Law." *Quarterly Review of Biology* **49** (March 1974): 3–26.

—., and Fowler, Ruth E. "Human Embryos in the Laboratory." *Scientific American* **233** (Ocotber 1970): 45-54.

—., and Sharpe, David J. "Social Values and Research in Human Embryology." *Nature* **231** (May 14, 1971): 87-91.

Fletcher, Joseph. *The Ethics of Genetic Control: Ending Reproductive Roulette.* Garden City, NY: Doubleday Anchor, 1974.

Francoeur, Robert T. *Utopian Motherhood.* New York: Doubleday, 1970.

Frankel, Mark S. *The Public Policy Dimensions of Artificial Insemination and Human Semen Cryobanking.* Program of Policy Studies in Science and Technology, Monograph No. 18. Washington, DC: George Washington University, December 1973.

Golding, Martin. "Ethical Issues in Biological Engineering." *UCLA Law Review* **15** (February, 1968): 443–479.

Hilton, Bruce, et al., eds. *Ethical Issues in Human Genetics: Genetic Counseling and the Use of Genetic Knowledge.* New York: Plenum Press, 1973.

Horne, Herbert W., Jr. "Artificial Insemination, Donor: An Issue of Ethical and Moral Values." *New England Journal of Medicine* **293** (Oct. 23, 1975): 873–874.

Kass, Leon R. "Babies by Means of In Vitro Fertilization: Unethical Experiments on the Unborn?" *New England Journal of Medicine* **285** (Nov. 18, 1971): 1174–1179.

—. "Making Babies: The New Biology and the 'Old' Morality." *Public Interest*, No. 26 (Winter, 1972): 18–56.

Lappe, Marc. "Moral Obligations and the Fallacies of Genetic Control." *Theological Studies* **33** (September, 1972): 411–427.

Law and Ethics of A.I.D.and Embryo Transfer. Ciba Foundation Symposium 17 (new series). New York: Associated Scientific Publishers, 1973.

Appendix

American Hospital Association

Patient's Bill of Rights (Chicago, 1970)

The patient has the right:

1. To considerate and respectful care
2. To obtain from the physician information regarding his or her diagnosis, treatment, and prognosis
3. To give informed consent before the start of any procedure or treatment
4. To refuse treatment to the extent permitted by law
5. To privacy concerning his or her own medical care program
6. To confidential communication and records
7. To expect that the hospital will make a reasonable response to a patient's request for service
8. To information regarding the relationship of his or her hospital to other health and educational institutions insofar as care is concerned
9. To refuse to participate in research projects
10. To expect reasonable continuity of care
11. To examine and question his or her bill
12. To know the hospital rules and regulations that apply to patients' conduct.

The Hippocratic Oath

I swear by Apollo Physician and Asclepius and Hygieia and Panaceia and all the gods and goddesses, making them my witnesses that I will fulfill according to my ability and judgment this oath and this covenant:

To hold him who has taught me this art as equal to my parents and to live my life in partnership with him, and if he is in need of money to give him a share of mine, and to regard his off-spring as equal to my brothers in male lineage and to teach them this art—if they desire to learn it—without fee and covenant; to give a share of precepts and oral instructions and all the other learning to my sons and to the sons of him who has instructed me and to pupils who have signed the covenant and have taken an oath according to the medical law, but to no one else.

I will apply dietetic measures for the benefit of the sick according to my ability and judgment; I will keep them from harm and injustice.

I will neither give a deadly drug to anybody if asked for it, nor will I make a suggestion to this effect. Similarly I will not give to a woman an abortive remedy. In purity and holiness I will guard my life and my art.

I will not use the knife, not even on sufferers from stone, but will withdraw in favor of such men as are engaged in this work.

Whatever houses I may visit, I will come for the benefit of the sick, remaining free of all intentional injustice, of all mischief, and in particular of sexual relations with both female and male persons, be they free or slaves.

What I may see or hear in the course of the treatment or even outside of the treatment in regard to the life of men, which on no account one must spread abroad, I will keep to myself holding such things shameful to be spoken about.

If I fulfill this oath and do not violate it, may it be granted to me to enjoy life and art, being honored with fame among all men for all time to come; if I transgress it and swear falsely, may the opposite of all this be true.

The Florence Nightingale Pledge

I solemnly pledge myself before God and in presence of this assembly; to pass my life in purity and to practice my profession faithfully. I will abstain from whatever is deleterious and mischievous and will not take or knowingly administer any harmful drug.

I will do all in my power to maintain and elevate the standard of my profession and will hold in confidence all personal matters committed to my keeping and family affairs coming to my knowledge in the practice of my calling.

With loyalty will I endeavor to aid the physician in his work, and devote myself to the welfare of those committed to my care.

American Nurses' Association Revised Code of Ethics (1976)

1. The nurse provides services with respect for human dignity and the uniqueness of the client unrestricted by considerations of social or economic status, personal attributes, or the nature of health problems.

2. The nurse safeguards the client's right to privacy by judiciously protecting information of a confidential nature.

3. The nurse acts to safeguard the client and the public when health care and safety are affected by the incompetent, unethical, or illegal practice of any person.

4. The nurse assumes responsibility and accountability for individual nursing judgments and actions.

5. The nurse maintains competence in nursing.

6. The nurse exercises informed judgment and uses individual competence and qualifications as criteria in seeking consultation, accepting responsibilities, and delegating nursing activities to others.

7. The nurse participates in activities that contribute to the ongoing development of the profession's body of knowledge.

8. The nurse participates in the profession's efforts to implement and improve standards of nursing.

9. The nurse participates in the profession's efforts to establish and maintain conditions of employment conducive to high quality nursing care.

10. The nurse participates in the profession's effort to protect the public from misinformation and misrepresentation and to maintain the integrity of nursing.

11. The nurse collaborates with members of the health professions and other citizens in promoting community and national efforts to meet the health needs of the public.

Principles of Medical Ethics*

Preamble

These principles are intended to aid physicians individually and collectively in maintaining a high level of ethical conduct. They are not laws but standards by which a physician may determine the propriety of his conduct in his relationship with patients, with colleagues, with members of allied professions, and with the public.

Section 1

The principal objective of the medical profession is to render service to humanity with full respect for the dignity of man. Physicians should merit the confidence of patients entrusted to their case, rendering to each a full measure of service and devotion.

Section 2

Physicians should strive continually to improve medical knowledge and skill, and should make available to their patients and colleagues the benefits of their professional attainments.

Section 3

A physician should practice a method of healing founded on a scientific basis; and he should not voluntarily associate professionally with anyone who violates this principle.

Section 4

The medical profession should safeguard the public and itself against physicians deficient in moral character or professional competence. Physicians should observe all laws, uphold the dignity and honor of the profession and accept its self-imposed disciplines. They should expose, without hesitation, illegal or unethical conduct of fellow members of the profession.

*Adopted by Judicial Council of AMA (1971)

Section 5

A physician may choose whom he will serve. In an emergency, however, he should render service to the best of his ability. Having undertaken the care of a patient, he may not neglect him; and unless he has been discharged he may discontinue his services only after giving adequate notice. He should not solicit patients.

Section 6

A physician should not dispose of his services under terms or conditions which tend to interfere with or impair the free and complete exercise of this medical judgment and skill or tend to cause a deterioration of the quality of medical care.

Section 7

In the practice of medicine a physician should limit the source of his professional income to medical services actually rendered by him, or under his supervision, to his patients. His fee should be commensurate with the services rendered and the patient's ability to pay. He should neither pay nor receive a commission for referral of patients. Drugs, remedies or appliances may be dispensed or supplied by the physician provided it is in the best interests of the patient.

Section 8

A physician should seek consultation upon request; in doubtful or difficult cases; or whenever it appears that the quality of medical services may be enhanced thereby.

Section 9

A physician may not reveal the confidences entrusted to him in the course of medical attendance, or the deficiencies he may observe in the character of patients, unless he is required to do so by law or unless it becomes necessary in order to protect the welfare of the individual or of the community.

Section 10

The honored ideals of the medical profession imply that the responsibilities of the physician extend not only to the individual, but also to society where these responsibilities deserve his interest and participation in activities which have the purpose of improving both the health and the well-being of the individual and the community.

Index